MODELING COUNT DATA

This definitive entry-level text, authored by a leading statistician in the field, offers clear and concise guidelines on how to select, construct, interpret, and evaluate count data. Written for researchers with little or no background in advanced statistics, the book presents treatments of all major models, using numerous tables, insets, and detailed modeling suggestions. It begins by demonstrating the fundamentals of modeling count data, including a thorough presentation of the Poisson model. It then works up to an analysis of the problem of overdispersion and of the negative binomial model, and finally to the many variations that can be made to the base count models. Examples in Stata, R, and SAS code enable readers to adapt models for their own purposes, making the text an ideal resource for researchers working in health, ecology, econometrics, transportation, and other fields.

Joseph M. Hilbe is a solar system ambassador with NASA's Jet Propulsion Laboratory, California Institute of Technology; an adjunct professor of statistics at Arizona State University; an emeritus professor at the University of Hawaii; and an instructor for Statistics.com, a web-based continuing-education program in statistics. He is currently president of the International Astrostatistics Association, and he is an elected Fellow of the American Statistical Association, for which he is the current chair of the section on Statistics in Sports. Author of several leading texts on statistical modeling, Hilbe also serves as the coordinating editor for the Cambridge University Press series Predictive Analytics in Action.

Other Statistics Books by Joseph M. Hilbe

Generalized Linear Models and Extensions (2001, 2007, 2013 – with J. Hardin)

Generalized Estimating Equations (2002, 2013 – with J. Hardin)

Negative Binomial Regression (2007, 2011)

Logistic Regression Models (2009)

Solutions Manual for Logistic Regression Models (2009)

R for Stata Users (2010 – with R. Muenchen)

Methods of Statistical Model Estimation (2013 – with A. Robinson)

A Beginner's Guide to GLM and GLMM with R: A Frequentist and Bayesian Perspective for Ecologists (2013 – with A. Zuur and E. Ieno)

Quasi–Least Squares Regression (2014 – with J. Shults)

Practical Predictive Analytics and Decisioning Systems for Medicine (2014 – with L. Miner, P. Bolding, M. Goldstein, T. Hill, R. Nisbit, N. Walton, and G. Miner)

MODELING COUNT DATA

JOSEPH M. HILBE

Arizona State University
and
Jet Propulsion Laboratory,
California Institute of Technology

CAMBRIDGE
UNIVERSITY PRESS

32 Avenue of the Americas, New York, NY 10013-2473, USA

Cambridge University Press is part of the University of Cambridge.

It furthers the University's mission by disseminating knowledge in the pursuit of education, learning, and research at the highest international levels of excellence.

www.cambridge.org
Information on this title: www.cambridge.org/9781107611252

First published 2014

Printed in the United States of America

A catalog record for this publication is available from the British Library.

ISBN 978-1-107-02833-3 Hardback
ISBN 978-1-107-61125-2 Paperback

Additional resources for this publication at www.cambridge.org/9781107611252

Contents

Preface

Modeling Count Data is written for the practicing researcher who has a reason to analyze and draw sound conclusions from modeling count data. More specifically, it is written for an analyst who needs to construct a count response model but is not sure how to proceed.

A count response model is a statistical model for which the dependent, or response, variable is a count. A count is understood as a nonnegative discrete integer ranging from zero to some specified greater number. This book aims to be a clear and understandable guide to the following points:

- How to recognize the characteristics of count data
- Understanding the assumptions on which a count model is based
- Determining whether data violate these assumptions (e.g., overdispersion), why this is so, and what can be done about it
- Selecting the most appropriate model for the data to be analyzed
- Constructing a well-fitted model
- Interpreting model parameters and associated statistics
- Predicting counts, rate ratios, and probabilities based on a model
- Evaluating the goodness-of-fit for each model discussed

There is indeed a lot to consider when selecting the best-fitted model for your data. I will do my best in these pages to clarify the foremost concepts and problems unique to modeling counts. If you follow along carefully, you should have a good overview of the subject and a basic working knowledge needed for constructing an appropriate model for your study data. I focus on understanding the nature of the most commonly used count models and on the problem of dealing with both over- and underdispersion, as well as on Poisson and negative binomial regression and their many variations. However, I also introduce several other count models that have not had much use in research because of the unavailability of commercial software for their estimation. In particular, I also discuss models such as the Poisson inverse Gaussian, generalized Poisson, varieties of three-parameter negative binomial, exact Poisson, and several other count models that will provide analysts with an expanded ability to better model the data at hand. Stata and/or R software and guidelines are provided for all of the models discussed in the text.

I am supposing that most people who will use this book start with little to no background in modeling count response data, although readers are expected to have a working knowledge of a major statistical software package, as well as a basic understanding of statistical regression. I provide an overview of maximum likelihood and iterative reweighted least squares (IRLS) regression in Sections 1.4.2 and 1.4.3, which assume an elementary understanding of calculus, but I consider these two sections as optional to our discussion. They are provided for those who are interested in how the majority of models we discuss are estimated. I recommend that you read these sections, even if you do not have the requisite mathematical background. I have attempted to present the material so that it will still be understood. Various terms are explained in these sections that will be used throughout the text.

Seasoned statisticians can also learn new material from the text, but I have specifically written it for researchers or analysts, as well as students at the upper-division to graduate levels, who want an entry-level book that focuses on the practical aspects of count modeling. The book is also addressed to statistical and predictive analytics consultants who find themselves faced with a project involving the modeling of count data, as well as to anyone with an interest in this class of statistical models. It is written in guidebook form, with lots of bullet points, tables, and complete statistical programming code for all examples discussed in the book.

Many readers of this book may be acquainted with my text *Negative Binomial Regression* (Cambridge University Press), which was first published in 2007. A substantially enhanced second edition was published in 2011. That text addresses nearly every count model for which there existed major statistical software support at the time of the book's publication. *Negative Binomial Regression* was primarily written for those who wish to understand the mathematics behind the models as well as the specifics and applications of each model. I recommend it for those who wish to go beyond the discussions found in *Modeling Count Data*.

I primarily use two statistical software packages to demonstrate examples of the count models discussed in the book. First, the Stata 13 statistical package (http://www.stata.com) is used throughout the text to display example model output. I show both Stata code and output for most of the modeling examples. I also provide R code (www.r-project.org) in the text that replicates, as far as possible, the Stata output. R output is also given when helpful. There are also times when no current Stata code exists for the modeling of a particular procedure. In such cases, R is used. SAS code for a number of the models discussed in the book is provided in the Appendix. SAS does not support many of the statistical functions and tests discussed later in the book, but its count-modeling capability is growing each year. I will advise readers on the book's web site as software for count models is developed for these packages. I should mention that I have used Stat/Transfer 12 (2013, Circle Systems) when converting data between statistical software packages. The user is able to convert between 37 different file formats, including those used in this book. It is a very helpful tool for those who must use more than one statistical or spreadsheet file.

Many of the Stata statistical models discussed in the text are offered as a standard part of the commercial package. Users have also contributed count model "commands" for the use of the greater Stata community. Developers of the user-authored commands used in the book are acknowledged at the first use of the software. James Hardin and I have both authored and coauthored a number of the more advanced count models found in the book. Many derive from our 2012 text *Generalized Linear Models and Extensions, 3rd edition* (Stata Press; Chapman & Hall/CRC). Several others in the book are based on commands we developed in 2013 for journal article publications. I should also mention that we also coauthored the current version of Stata's **glm** command (2001), although Stata has subsequently enhanced various options over the past 12 years as new versions of Stata were released. Several of the R functions and scripts used in the book were coauthored by Andrew Robinson and me for use in our book (Hilbe and Robinson 2013). Data sets and functions for this book, as well as for Hilbe (2011), are available in the **COUNT** package, which may be downloaded from any **CRAN** mirror site. I also recommend installing **msme** (Hilbe and Robinson), also available on **CRAN**. I have also posted all of my user-authored Stata commands and functions, as well as all data sets used in the book, on the book's web site at the following address: http://works.bepress.com/joseph_hilbe/. The book's page with Cambridge University Press is at www.cambridge.org/9781107611252.

The data files used for examples in the book are real data. The **rwm1984** and **medpar** data sets are used extensively throughout the book. Other data sets used include **titanic**, **heart**, **azcabgptca**, **smoking**, **fishing**, **fasttrakg**, **rwm5yr**, **nuts**, and **azprocedure**. The data are defined where first used. The **medpar**, **rwm5yr**, and **titanic** data are used more than other data in the book. The **medpar** data are from the 1991 Arizona Medicare files for two diagnostic groups related to cardiovascular procedures. I prepared **medpar** in 1993 for use in workshops I gave at the time. The **rwm5yr** data consist of 19,609 observations from the German Health Reform data covering the five-year period of 1984–1988. Not all patients were in the study for all five years. The count response is the number of visits made by a patient to the doctor during that calendar year. The **rwm1984** data were created from **rwm5yr**, with only data from 1984 included – one patient, one observation. The well-known **titanic** data set is from the 1912 *Titanic* ship disaster survival data. It is in grouped format with *survived* as the response. The predictors are *age* (adult vs. child), *gender* (male vs. female), and *class* (1st-, 2nd-, and 3rd-class passengers). Crew members have been excluded.

I advise the reader that there are parts of Chapter 3 that use or adapt text from the first edition of *Negative Binomial Regression* (Hilbe 2007a), which is now out of print, as it was superseded by Hilbe (2011). Chapter 2 incorporates two tables that were also used in the first edition. I received very good feedback regarding these sections and found no reason to change them for this book. Now that the original book is out of print, these sections would be otherwise lost.

I wish to acknowledge five eminent colleagues and friends in the truest sense who in various ways have substantially contributed to this book, either indirectly while working together on other projects or directly: James Hardin, director of the Biostatistics Collaborative Unit and professor, Department of Statistics and Epidemiology, University of South Carolina School of Medicine; Andrew Robinson, director, Australian Centre of Excellence for Risk Analysis (ACERA), Department of Mathematics and Statistics, University of Melbourne, Australia; Alain Zuur, senior statistician and director of Highland Statistics Ltd., UK; Peter Bruce, CEO, Institute for Statistics Education (Statistics.com); and John Nelder, late Emeritus Professor of Statistics, Imperial College, UK. John passed away in 2010, just shy of his eighty-sixth birthday; our many discussions over a 20-year period are sorely missed. He definitely spurred my interest in the negative binomial model. I am fortunate to have known and to have worked with these fine statisticians. Each has enriched my life in different ways.

Others who have contributed to this book's creation include Valerie Troiano and Kuber Dekar of the Institute for Statistics Education; Professor William H. Greene, Department of Economics, New York University, and author of the Limdep econometrics software; Dr. Gordon Johnston, Senior Statistician, SAS Institute, author of the SAS Genmod Procedure; Professor Milan Hejtmanek, Seoul National University, and Dr. Digant Gupta, M.D., director, Outcomes Research, Cancer Treatment Centers of America, both of whom provided long hours reviewing early drafts of the book manuscript. Helen Wheeler, production editor for Cambridge University Press, is also gratefully acknowledged. A special acknowledgment goes to Patricia Branton of Stata Corp., who has provided me with statistical support and friendship for almost a quarter of a century. She has been a part of nearly every text I have written on statistical modeling, including this book.

There have been many others who have contributed to this book as well, but space limits their express acknowledgment. I intend to list all contributors on the book's web site. I invite readers to contact me regarding comments or suggestions about the book. You may email me at hilbe@asu.edu or at the address on my BePress web site listed earlier.

Finally, I must also acknowledge Diana Gillooly, senior editor for mathematical sciences with Cambridge University Press, who first encouraged me to write this monograph. She has provided me with excellent feedback in my attempt to develop a thoroughly applied book on count models. Her help with this book has been invaluable and goes far beyond standard editorial obligations. I also wish to thank my family for yet again supporting my writing of another book. My appreciation goes to my wife, Cheryl L. Hilbe, my children and grandchildren, and our white Maltese dog, Sirr, who sits close by my side for hours while I am typing. I dedicate this book to Cheryl for her support and feedback during the time of this book's preparation.

Joseph M. Hilbe
Florence, Arizona
August 12, 2013

## CHAPTER 1

Varieties of Count Data

SOME POINTS OF DISCUSSION

- What are counts? What are count data?
- What is a linear statistical model?
- What is the relationship between a probability distribution function (PDF) and a statistical model?
- What are the parameters of a statistical model? Where do they come from, and can we ever truly know them?
- How does a count model differ from other regression models?
- What are the basic count models, and how do they relate with one another?
- What is overdispersion, and why is it considered to be the fundamental problem when modeling count data?

1.1 WHAT ARE COUNTS?

When discussing the modeling of count data, it's important to clarify exactly what is meant by a count, as well as "count data" and "count variable." The word "count" is typically used as a verb meaning to enumerate units, items, or events. We might count the *number* of road kills observed on a stretch of highway, *how many* patients died at a particular hospital within 48 hours of having a myocardial infarction, or *how many* separate sunspots were observed in March 2013. "Count data," on the other hand, is a plural noun referring to observations made about events or items that are enumerated. In statistics, count data refer to observations that have only nonnegative integer values ranging from zero to some greater undetermined value. Theoretically, counts can range from zero to infinity, but they are always limited to some lesser distinct value – generally the maximum value of the count data being modeled. When the data being modeled consist of a large number of distinct values, even if they are positive integers, many statisticians prefer to model the counts as if they were continuous data. We address this issue later in the book.

A "count variable" is a specific list or array of count data. Again, such observations can only take on nonnegative integer values. However, in a statistical model, a response variable is understood as being a random variable, meaning that the particular set of enumerated values or counts could be other than they are at any given time. Moreover, the values are assumed to be independent of one another (i.e., they show no clear evidence of correlation). This is an important criterion for count model data, and it stems from the fact that the observations of a probability distribution are independent. On the other hand, predictor values are fixed; that is, they are given as facts, which are used to better understand the response.

We will be primarily concerned with four types of count variables in this book. They are:

1. A count or enumeration of events
2. A count of items or events occurring within a period of time or over a number of periods
3. A count of items or events occurring in a given geographical or spatial area or over various defined areas
4. A count of the number of people having a particular disease, adjusted by the size of the population at risk of contracting the disease

Understanding how count data are modeled, and what modeling entails, is discussed in the following section. For readers with little background in linear models, I strongly suggest that you read through Chapter 1 even though various points may not be fully understood. Then re-read the chapter carefully. The essential concepts and relationships involved in modeling should then be clear. In Chapter 1, I have presented the fundamentals of modeling, focusing on normal and count model estimation from several viewpoints, which should at the end provide the reader with a sense of how the modeling process is to be understood when applied to count models. If certain points are still unclear, I am confident that any problem areas regarding the assessment of fit will be clear by the time you read through Chapter 4, on assessing model fit. Those who have taken a statistics course in which linear regression is examined should have no problem following the presentation.

1.2 UNDERSTANDING A STATISTICAL COUNT MODEL

1.2.1 Basic Structure of a Linear Statistical Model

Statistics may be generically understood as the science of collecting and analyzing data for the purpose of classification, prediction, and of attempting to quantify and understand the uncertainty inherent in phenomena underlying data.

A statistical model describes the relationship between one or more variables on the basis of another variable or variables. For the purpose of the models we discuss in this book, a statistical model can be understood as the mathematical explanation of a count variable on the basis of one or more explanatory variables.[1] Such statistical models are stochastic, meaning that they are based on probability functions. The traditional linear regression model is based on the normal or Gaussian probability distribution and can be formalized in the most simple case as

$$Y = \beta_0 + \beta X + \varepsilon \tag{1.1}$$

where Y is called the response, outcome, dependent, or sometimes just the y variable. We use the term "response" or y when referring to the variable being modeled. X is the explanatory or predictor variable that is used to explain the occurrence of y. β is the coefficient for X. It is a slope describing the rate of change in the response based on a one-unit change in X, holding other predictor values constant (usually at their mean values). β_0 is the intercept, which provides a value to fitted y, or $\hat{y}$, when, or if, X has the value of 0. ε (eta) is the error term, which reflects the fact that the relationship between X and Y is not exact, or deterministic. For the normal or linear regression model, the errors are Gaussian or normally distributed, which is the most

---

[1] A model may consist of only the response variable, unadjusted by explanatory variables. Such a model is estimated by modeling the response on the intercept. For example, using R: *lm(y ~ 1)*; using Stata: *reg y*.

well-used and basic probability distribution in statistics. ε is also referred to as the residual term.

When a linear regression has more than one predictor, it may be schematized by giving a separate *beta* and X value for each predictor, as

$$Y = \beta_0 + \beta_1 X_1 + \beta_2 X_2 + \cdots + \beta_n X_n + \varepsilon \tag{1.2}$$

Statisticians usually convert equation (1.2) to one that has the left-hand side being the predicted or expected mean value of the response, based on the sum of the predictors and coefficients. Each associated coefficient and predictor is called a regression *term*:

$$\hat{y} = \beta_0 + \beta_1 X_1 + \beta_2 X_2 + \cdots + \beta_n X_n \tag{1.3}$$

or

$$\hat{\mu} = \beta_0 + \beta_1 X_1 + \beta_2 X_2 + \cdots + \beta_n X_n \tag{1.4}$$

Notice that the error became part of the expected or predicted mean response. "$\hat{}$", or *hat* over y and μ (*mu*), indicates that this is an estimated value. From this point on, I use the symbol μ to refer to the predicted value, without a *hat*. Understand, though, that when we are estimating a parameter or a statistic, a *hat* should go over it. The true unknown parameter, on the other hand, has no *hat*. You will also at times see the term $E(y)$ used to mean "estimated y." I will not use it here.

In matrix form, where the individual terms of the regression are expressed in a single term, we have

$$\mu = \beta X \tag{1.5}$$

with βX being understood as the summation of the various terms, including the intercept. As you may recall, the intercept is defined as $\beta_0(1)$, or simply β_0. It is therefore a term that can be placed within the single matrix term βX. When models become complicated, viewing them in matrix form is the only feasible way to see the various relationships involved. I should mention that sometimes you see the term βX expressed as $x\beta$. I reserve this symbol for another part of the model, which we discuss a bit later in this section.

Let's look at example data (**smoking**). Suppose that we have a six-observation model consisting of the following variables:

sbp: systolic blood pressure of subject
male: 1 = male; 0 = female
smoker: 1 = history of smoking; 0 = no history of smoking
age: age of subject

Using Stata statistical software, we display a linear regression of *sbp* on *male*, *smoker*, and *age*, producing the following (*nohead* suppresses the display of header statistics).

```
STATA CODE
. regress sbp male smoker age, nohead
---------------------------------------------------------------------
    sbp |      Coef.   Std. Err.      t    P>|t|     [95% Conf. Interval]
--------+------------------------------------------------------------
   male |   4.048601   .2507664    16.14   0.004      2.96964    5.127562
 smoker |   6.927835   .1946711    35.59   0.001     6.090233    7.765437
    age |   .4698085    .02886     16.28   0.004     .3456341     .593983
  _cons |   104.0059   .7751557   134.17   0.000     100.6707    107.3411
---------------------------------------------------------------------
```

Continuing with Stata, we may obtain the predicted value, μ, which is the estimated mean systolic blood pressure, and display the predictor values together with μ (mu) as

```
. predict mu
. l                    // 'l' is an abbreviation for list
     +------------------------------------+
     |  sbp   male   smoker   sge      mu  |
     |------------------------------------|
  1. |  131     1      1        34   130.9558 |
  2. |  132     1      1        36   131.8954 |
  3. |  122     1      0        30   122.1488 |
  4. |  119     0      0        32   119.0398 |
  5. |  123     0      1        26   123.1488 |
  6. |  115     0      0        23   114.8115 |
     +------------------------------------+
```

To see exactly what this means, we sum the terms of the regression. The intercept term is also summed, but its values are set at 1. The _b[] term captures the coefficient from the results saved by the software. For the intercept, _b[_cons] adds the intercept term, slope[1], to the other values. The term *xb* is also commonly referred to as the *linear predictor*.

```
. gen xb =  _b[male]*male + _b[smoker]*smoker + _b[age]*age + _b[_cons]
. l

   +---------------------------------------------------+
   | sbp    male    smoker   age      mu         xb    |
   |---------------------------------------------------|
1. | 131     1       1       34    130.9558   130.9558 |
2. | 132     1       1       36    131.8954   131.8954 |
3. | 122     1       0       30    122.1488   122.1488 |
4. | 119     0       0       32    119.0398   119.0398 |
5. | 123     0       1       26    123.1488   123.1488 |
6. | 115     0       0       23    114.8115   114.8115 |
   +---------------------------------------------------+
```

The intercept is defined correctly; check by displaying it. The value is indeed 1,

```
. di _cons
1
```

whereas _b[_cons] is the constant slope of the intercept as given in the preceding regression output:

```
. di _b[_cons]    /* intercept slope */
104.00589
```

Using R, we may obtain the same results with the following code:

```
R CODE
> sbp    <- c(131,132,122,119,123,115)
> male   <- c(1,1,1,0,0,0)
> smoker <- c(1,1,0,0,1,0)
> age    <- c(34,36,30,32,26,23)
> summary(reg1 <- lm(sbp~ male+smoker+age))
         <results not displayed>
```

Predicted values may be obtained by

```
> mu <- predict(reg1)
> mu
        1        2        3        4        5        6
 130.9558 131.8954 122.1487 119.0398 123.1487 114.8115
```

As was done with the Stata code, we may calculate the linear predictor, which is the same as μ, by first abstracting the coefficient

```
> cof <- reg1$coef
> cof
(Intercept)         male      smoker           age
104.0058910    4.0486009   6.9278351     0.4698085
```

and then the linear predictor, *xb*. Each coefficient can be identified with [ ]. The values are identical to *mu*.

```
> xb <- cof[1] + cof[2]*male + cof[3]*smoker + cof[4]*age
> xb
[1] 130.9558 131.8954 122.1487 119.0398 123.1487 114.8115
```

Notice the closeness of the observed response and predicted values. The differences are

```
> diff <- sbp - mu
> diff
            1            2            3            4            5            6
   0.04418262   0.10456554  -0.14874816  -0.03976436  -0.14874816   0.18851252
```

When the values of the linear predictor are close to the predicted or expected values, we call the model *well fitted*.

1.2.2 Models and Probability

One of the points about statistical modeling rarely discussed is the relationship of the data to a probability distribution. All parametric statistical models are based on an underlying probability distribution. I mentioned before that the normal or linear regression model is based on the Gaussian, or normal, probability distribution (see example in Figure 1.1). It is what defines the error terms. When we are attempting to estimate a least squares regression or more sophisticated maximum likelihood model, we are estimating the parameters of the underlying probability distribution that characterize the data. These two foremost methods of estimation are described in the next section of this opening chapter. The important point here is always to remember that when modeling count data we are really estimating the parameters of a probability distribution that we believe best represents the data we are modeling. We are never able to knowingly determine the true parameters

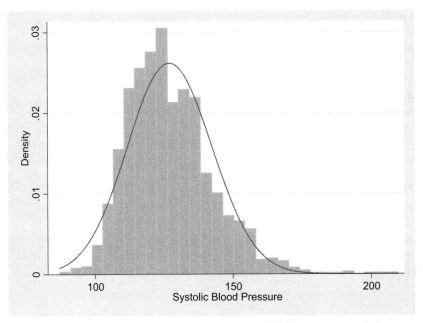

FIGURE 1.1. Gaussian distribution approximated by blood pressure data.

of the probability distribution function, which we shall refer to as the PDF, but we attempt to obtain the best unbiased estimate possible. The parameters are what provide the shape of the PDF we are using to describe the data. By knowing the estimated parameter or parameters, we can use them to predict data from inside the sample of data from which we are modeling and in special cases data from outside the sample.

This is also an important point to keep in mind. We assume that the data being modeled are a random sample from a greater population of data. The PDF whose parameters we are attempting to estimate is assumed to describe the population data, not only the sample from it that we are actually modeling. This way of looking at statistics and data is commonly referred to as frequency-based statistical modeling. Bayesian models look at the relationship of data to probability distributions in a different manner, which we discuss in the final chapter. However, the standard way of modeling is based on this frequency interpretation, which was championed by Ronald Fisher in the early twentieth century and has dominated statistics since. I might say here, though, that many statisticians are turning to Bayesian estimation when modeling certain types of data. Again, we'll address this situation in the final chapter, proposing several predictions in the process.

1.2.3 Count Models

The majority of count models discussed in this book are based on two probability distributions – the Poisson and negative binomial PDFs. I add three additional models in this volume that I consider important when initially evaluating count data – the Poisson inverse Gaussian model, or PIG, Greene's three-parameter negative binomial P, or NB-P, and generalized Poisson (GP) models. These five distributions are closely related. The Poisson distribution has a single parameter to be estimated, μ, or the mean, which is also sometimes referred to as the location parameter. The unique feature of the Poisson distribution is that the mean and variance are the same. The higher the value of the mean of the distribution, the greater the variance or variability in the data. For instance, if we are modeling the number of cars failing to properly stop at two different stop signs per day over a period of a month, and if the average number of failures to stop per day at Site A is 4 and at Site B is 8, we automatically know that the variance of the distribution of failures at Site A is also 4 and at Site B is 8. No other measurements need be done – that is, if the true distribution at each site is Poisson. Recall from algebra that the variance is the square of the standard deviation. The mean and standard deviation of the counts of failures at Site A are 4 and 2, respectively, and at Site B are 8 and $2\sqrt{2}$.

This criterion of the Poisson distribution is referred to as the equidispersion criterion. The problem is that when modeling real data, the equidispersion criterion is rarely satisfied. Analysts usually must adjust their Poisson model in some way to account for any under- or overdispersion that is in the data. Overdispersion is by far the foremost problem facing analysts who use Poisson regression when modeling count data.

I should be clear about the meaning of overdispersion since it is central to the modeling of count data and therefore plays an important role in this book. Overdispersion almost always refers to excess variability or correlation in a Poisson model, but it also needs to be considered when modeling other count models as well. Keep in mind, however, that when the term "overdispersion" is used, most analysts are referring to Poisson overdispersion (i.e., overdispersion in a Poisson model).

Simply put, Poisson overdispersion occurs in data where the variability of the data is greater than the mean. Overdispersion also is used to describe data in a slightly more general sense, as when the observed or "in fact" variance of the count response is greater than the variance of the predicted or expected counts. This latter type of variance is called expected variance. Again, if the observed variance of the response is greater than the expected variance, the data are overdispersed. A model that fails to properly adjust for overdispersed data is called an overdispersed model. As such, its standard errors are biased and cannot be trusted. The standard errors associated with model predictors may appear from the model to significantly contribute to the understanding of the response, but in fact they may not. Many analysts have been deceived into thinking that they have developed a well-fitted model.

Unfortunately, statistical software at times fails to provide an analyst with the information needed to determine if a Poisson model is overdispersed or underdispersed. We discuss in some detail exactly how we can determine whether a model is overdispersed. More properly perhaps, this book will provide guidelines to help you decide whether a Poisson model is equidispersed.

Probably the most popular method of dealing with apparent Poisson overdispersion is to model the data using a negative binomial model. The negative binomial distribution has an extra parameter, referred to as the negative binomial dispersion parameter. Some books and articles call the dispersion parameter the *heterogeneity* parameter or *ancillary* parameter. These are appropriate names as well. The dispersion parameter is a measure of the adjustment needed to accommodate the extra variability, or heterogeneity, in the data. However, the term *dispersion* parameter has become the standard name for the second parameter of the negative binomial distribution.

The negative binomial, which we discuss in more detail later, allows more flexibility in modeling overdispersed data than does a single-parameter Poisson model. The negative binomial is derived as a Poisson-gamma mixture model, with the dispersion parameter being distributed as gamma shaped. The gamma PDF is pliable and allows for a wide variety of shapes. As a consequence, most overdispersed count data can be appropriately modeled using a negative binomial regression. The advantage of using the negative binomial rests with the fact that when the dispersion parameter is zero (0), the model is Poisson.[2] Values of the dispersion parameter greater than zero indicate that the model has adjusted for correspondingly greater amounts of

---

[2] I term this the direct parameterization of the negative binomial. Unlike most commercial statistical software, R's **glm** and **glm.nb** functions employ an inverted relationship of the dispersion parameter, theta, so that a Poisson model results when theta approaches infinity. Most subsequent R functions have followed **glm** and **glm.nb**. I maintain the direct relationship for all count models in this volume and discuss the differences between the two parameterizations in some detail later in the book.

overdispersion. The negative binomial dispersion parameter will be symbolized as α (*alpha*).

We will later discover that there are a variety of reasons why data can have more variability than what is assumed based on the Poisson distribution. For that matter, the count response variable of a negative binomial model can also have more variability than allowed by negative binomial distributional assumptions. In that sense, a negative binomial model may also be over- or underdispersed. It is important to remember, though, and this point will be repeated later, that the negative binomial model cannot be used to adjust for Poisson underdispersion. That is:

> *The negative binomial model adjusts for Poisson overdispersion; it cannot be used to model underdispersed Poisson data.*

Statisticians recognize an alternative linear negative binomial model commonly referred to as NB1, which differs from the standard model described previously, also known as NB2. I delay discussion of the NB1 model until later in the book.

The third type of count model that I wish to introduce in this book is the Poisson inverse Gaussian model, also fondly known as the PIG model. The PIG model assumes that overdispersion in a Poisson model is best described, or shaped, according to the inverse Gaussian distribution rather than the gamma distribution that is inherent to the negative binomial model. The reason that the PIG model has not had wide use is that until very recently there was no software for its estimation – except if analysts created the software themselves. Programming a PIG is not easy, and unless programmed well, a PIG algorithm can take a long time to converge with appropriate parameter estimates. The **gamlss** R package (Rigby and Stasinopoulos 2008) provides support for PIG modeling and may be downloaded from CRAN. The **pigreg** command (Hardin and Hilbe 2012) is available for Stata users and may be downloaded from the author's web site or from the publisher's web sites for Hilbe (2011) or Hardin and Hilbe (2012). The PIG dispersion parameter in **pigreg** is also known as α in that it has an interpretation similar to that of the negative binomial dispersion parameter. Like the negative binomial, we parameterized the PIG α parameter so that a value of $\alpha = 0$ is Poisson.

The fourth type of count model provided in this book is a three-parameter generalized negative binomial model designed by William Greene of New York University, who is the author of the Limdep econometric statistical

TABLE 1.1. Selected Count Model Mean–Variance Relationship		
Model	Mean	Variance
Poisson	μ	μ
Negative binomial (NB1)	μ	$\mu(1 + \alpha) = \mu + \alpha\mu$
Negative binomial (NB2)	μ	$\mu(1 + \alpha\mu) = \mu + \alpha\mu^2$
Poisson inverse Gaussian	μ	$\mu(1 + \alpha\mu^2) = \mu + \alpha\mu^3$
Negative binomial-P	μ	$\mu(1 + \alpha\mu^\rho) = \mu + \alpha\mu^\rho$
Generalized Poisson	μ	$\mu(1 + \alpha\mu)^2 = \mu + 2\alpha\mu^3 + \alpha^2\mu^3$

software program. It is called NB-P. The dispersion parameter, α, of negative binomial and PIG models has the same value for all observations in the model. The third parameter of NB-P, called ρ or rho in Greek, allows the dispersion to vary across observations, providing a better opportunity to fit negative binomial data. The underlying PDF is still a variety of negative binomial and is usually used to help analysts decide on employing an NB1 or NB2 model on their data. We discuss how best to implement NB-P in Section 5.4.

The fifth model that I believe to be important when first considering how to model count data is the generalized Poisson (GP) model. Similar to the negative binomial and PIG models, the generalized Poisson has a second parameter, also referred to as the dispersion or scale parameter. Also like the previously introduced models, the generalized Poisson reduces to Poisson when the dispersion is zero. The nice feature of the generalized Poisson, however, is that the dispersion parameter can have negative values, which indicate an adjustment for Poisson underdispersion.

The mean and variance functions of the Poisson model and the five models just discussed are given in Table 1.1.

R and Stata code to graph the above five distributions with a given mean of 4 and, except for the Poisson, a dispersion parameter value of 0.5 is provided in Tables 1.2a and 1.2b, respectively. The figures produced by the code are displayed as Figures 1.2a and 1.2b.

Note how sharply peaked the PIG distribution is at the beginning of the distribution. This same general shape is the same for PIG regardless of the value of *alpha* or the mean.

The Stata graphic of the same distributions appears in Figure 1.2b. Note that the top of the PIG distribution is truncated at about 1.0. Care must be

TABLE 1.2A. R Code for Figure 1.2a

```
obs <- 15; mu <- 4; y <- (0:140)/10; alpha <- .5
amu <- mu*alpha; layout(1)
all.lines <- vector(mode = 'list', length = 5)
for (i in 1:length(mu)) {
    yp = exp(-mu[i])*(mu[i]^y)/factorial(y)
    ynb1 = exp( log(gamma(mu[i]/alpha + y))
                - log(gamma(y+1))
                - log(gamma(mu[i]/alpha))
                + (mu[i]/alpha)*log(1/(1+alpha))
                + y*log(1-1/(1+alpha)))
    ynb2 = exp( y*log(amu[i]/(1+amu[i]))
                - (1/alpha)*log(1+amu[i])
                + log( gamma(y +1/alpha) )
                - log( gamma(y+1) )
                - log( gamma(1/alpha) ))
    ypig = exp( (-(y-mu[i])^2)/(alpha*2*y*mu[i]^2))
                * (sqrt(1/(alpha*2*pi*y^3))))
    ygp = exp( log((1-alpha)*mu[i])
               + (y-1)*log((1-alpha) * mu[i]+alpha*y)
               - (1-alpha)*mu[i]
               - alpha*y
               - log(gamma(y+1)))
    all.lines = list(yp = yp, ynb1 = ynb1, ynb2 = ynb2, ypig = ypig,
    ygp = ygp)
    ymax = max(unlist(all.lines), na.rm=TRUE)
    cols = c("red","blue","black","green","purple")
    plot(y, all.lines[[1]], ylim =
        c(0, ymax), type = "n", main="5 Count Distributions:
        mean=4; alpha=0.5")
    for (j in 1:5)
        lines(y, all.lines[[j]], ylim = c(0, ymax),
        col=cols[j], type='b',pch=19, lty=j)
        legend("topright",cex = 1.5, pch=19,
        legend=c("NB2","POI","PIG","NB1","GP"),
        col = c(1,2,3,4,5),
        lty = c(1,1,1,1,1),
        lwd = c(1,1,1,1,3))
}
```

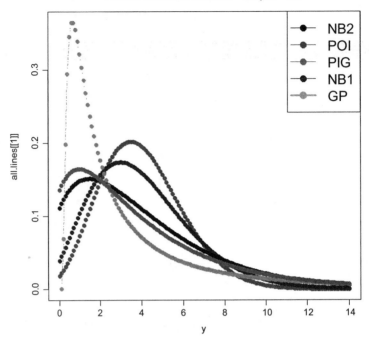

5 Count Distributions: mean = 4; alpha = 0.5

FIGURE 1.2A. Five count distributions: mean = 4; alpha = 0.5.

TABLE 1.2B. Stata Code for Figure 1.2b

```
clear
set obs 15
gen byte mu = 4
gen byte y = _n-1
gen yp = (exp(-mu)*mu^y)/exp(lngamma(y+1))
gen alpha = .5
gen amu = mu*alpha
gen ynb = exp(y*ln(amu/(1+amu)) - (1/alpha)*ln(1+amu) + lngamma(y +1/
          alpha) /*
 */ - lngamma(y+1) - lngamma(1/alpha))
gen ypig = exp((-(y-mu)^2)/(alpha*2*y*mu^2)) * (sqrt(1/(alpha*2*_pi
          *y^3)))
gen ygp = exp(ln((1-alpha)*mu) + (y-1)*ln((1-alpha)*mu+alpha*y) - /*
 */ (1-alpha)*mu - alpha*y - lngamma(y+1))
gen ynb1 = exp(lngamma(mu/alpha + y) - lngamma(y+1)- lngamma(mu/
          alpha) + /*
 */ (mu/alpha)*ln(1/(1+alpha)) + y*ln(1-1/(1+alpha)))
lab var yp "Poisson"
lab var ynb "NB2"
lab var ynb1 "NB1"
lab var ypig "PIG"
lab var ygp "generalized Poisson"
graph twoway connected ygp ypig ynb1 ynb yp y, ms(s D T d S) /*
 */ title("POI | NB | NB1 | GP |PIG distributions: MEAN = 4; a=0.5")
```

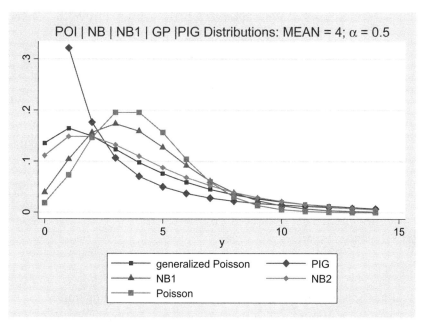

FIGURE 1.2B. Stata graphic of the same distributions as in Figure 1.2a.

taken when interpreting figures such as these to assure that the entire distribution is being displayed or that a notation is provided when the distribution appears to be truncated but is not. I advise checking the actual range of values that are expected for a given distribution.

A sixth type of count model that will prove to be of considerable value when modeling count data is the heterogeneous negative binomial, or NBH. The heterogeneous negative binomial allows for the parameterization of the dispersion parameter. That is, the dispersion parameter will have associated predictor coefficients indicating which ones significantly contribute to the overdispersion in the model. This feature of the NBH assists the analyst in determining the possible source for both Poisson and negative binomial overdispersion.

Other models will be discussed in the book as well, including truncated and zero-inflated models, hurdle models, other generalized models with three parameters, panel models, quantile models, exact Poisson models, Bayesian models, and others. However, the Poisson, negative binomial, and PIG models will play primary roles when we first attempt to identify the model that best fits the count data to be analyzed. Those models are considered in Chapters 2, 5, and 6. We elaborate on the nature of overdispersion in Chapter 3.

1.2.4 Structure of a Count Model

Nearly all of the count models we review have the basic structure of the linear model described earlier in this section. The difference is that the left-hand side of the equation is in log form:

$$\ln(\mu) = \beta_0 + \beta_1 X_1 + \beta_2 X_2 + \cdots + \beta_n X_n \tag{1.6}$$

To isolate the predicted mean count on the left side of equation (1.6), both sides of the equation are exponentiated, giving

$$\mu = e^{\beta_0 + \beta_1 X_1 + \beta_2 X_2 + \cdots + \beta_n X_n} \tag{1.7}$$

We will later discover that both of these expressions will be important when defining terms in count models. Notice that there is not a linear relationship between μ and the predictors as there was for the linear model. The linear relationship is between the natural log of μ and the predictors.

Recall that earlier in this section I referred to the sum of regression terms as the linear predictor. In the case of linear regression, the linear predictor is the same as the predicted or expected value. Statisticians typically symbolize the summation of the terms of the linear predictor for each observation in a model as

$$(x\beta)_i = \sum_{i=1}^{n} \beta_0 + \beta_1 X_{1i} + \cdots + \beta_j X_{ji} \tag{1.8}$$

with i indicating the observation number in the model data and j the number of predictors in the model. Notice that I used the standard mathematical Σ (Sigma) symbol for summation in equation (1.8). The summation starts at the quantity indicated below Sigma and ends with the value at its top. Here we have observation number i, starting at 1 representing the first observation in the data, and finishing with n, indicating the last observation in the data being modeled. At times, statisticians choose not to use the summation sign, or product sign for probabilities and likelihoods, or to use subscripts if they believe that their readers understand that this is included in the equation even if not displayed. Subscripts and summation symbols are generally not displayed if we are referring to the equation as a whole and not to the individual components. Each author has their individual preference.

The relationship of the predicted or fitted statistic, μ, and the linear predictor, xb, is the same for Poisson, negative binomial, and PIG regressions.[3] The term $\log(\mu)$ is called the link function since it links the linear predictor and predicted value:

$$\log(\mu_i) = \sum_{i=1}^{n} \beta_0 + \beta_1 X_{1i} + \cdots + \beta_j X_{ji} \qquad (1.9)$$

To summarize, the link between μ and xb is

```
REGRESSION           LINK RELATIONSHIP     PDF
----------------------------------------------------------------------
linear regression    identity: xb = μ      Gaussian or normal
major count models   log:      xb = ln(μ)  Poisson, negative binomial, PIG
----------------------------------------------------------------------
```

An important feature of having the natural log link for count models is that it guarantees that the predicted values will always be positive (i.e., $\mu > 0$). Using a linear regression when modeling counts cannot make such a guarantee. What about the formerly common practice of logging the response and modeling it as a linear regression (i.e., as a Gaussian model)? The result is that predicted values are positive but substantially lower than Poisson or negative binomial predicted values. Ideally, the predicted counts should be close to the values of the actual counts: $y \sim \mu$. When the count response is logged and modeled using linear regression, its predicted values are nearly always distant from the actual or observed counts. Fit tests that match up observed versus expected counts rarely show a significant fit. The caveat here is

> *Reject the temptation to use linear regression to model a logged count variable.*

At the very start of the modeling process, it's imperative to become intimately familiar with the data you intend to model. Given that the count models discussed in this book have a single response variable that is being modeled, it

---

[3] Given the way that the NB and PIG models have been parameterized here.

TABLE 1.3. Points to Understand about Your Model
• What is the response or dependent variable? Which variable(s) are we trying to understand on the basis of other variables?
• What type of values characterize the response variable? What types of values are characteristic of the explanatory predictors?
• How do the predictor variables relate to one another? Are there interaction effects or excessive correlation between them?
• What are the optimal predictors to use in order to best explain the response?
• How is the response variable distributed?
• Are there missing values in the model? If so, how are they distributed in the data?

is important to understand the points given in Table 1.3 about your proposed model.

These are just the foremost items that you need to know about the data you intend to model. Aside from the data itself, though, we also must know the context of the model we wish to develop. What's the point? This is perhaps a more important criterion than is at first realized. But as you will find when traversing the book, understanding the purpose of the model, or the purpose of the research, can help greatly when determining which predictors to use in the model. It's not just a matter of discarding all predictors with p-values greater than 0.05. It may be important to retain certain predictors in a model even though they do not appear to contribute significantly to an understanding of the response.

1.3 VARIETIES OF COUNT MODELS

I don't want to mislead you into thinking that the count response variable being modeled, as perhaps adjusted by various explanatory predictors, actually comes from some probability generating function or probability function. Not usually! The counts typically come from observations we make from the world around us, or even perhaps as a result of some study design. However, as far as a statistical model is concerned, the data can be thought of as the realization of a process that resembles the data being modeled. If we can estimate the parameters of the distribution underlying the data with as little bias as possible, we can use our understanding of probability to predict observations of real events outside the model; for example, future events. We use the underlying probability function and its estimated parameter(s) to make predictions and classifications about real data – data that belong to the population of data from which the model data are considered to be a sample.

If this characterization of the relationship of probability and model data is unclear now, I am confident that it will be clear to you by the end of the book. It is an important distinction that is basic to the frequency-based interpretation of statistics. Essentially, in selecting the most appropriate model for given data, the analyst is selecting a probability distribution, or mixtures of probability distributions, that best describe the population data of which the data being modeled are a random sample. I have mentioned this before, but it is an extremely important fact to understand about modeling.

As indicated in the last section, there are three primary probability functions that will initially concern us in this book: the Poisson, negative binomial, and Poisson inverse Gaussian models. Many of the other models we discuss are variations of these models – in particular, variations of the Poisson and negative binomial models.

Data come to us in a variety of ways. They certainly do not always perfectly match the Poisson or negative binomial PDF. For example, the Poisson, negative binomial, and PIG distributions each assume the possibility of zero counts even if there may not in fact be any. If zero counts are not a possibility for the data being modeled, for instance hospital length-of-stay data, then the underlying PDF may need to be amended to adjust for the excluded zero counts. Zero-truncated (ZT) models are constructed for exactly that purpose. The underlying PDF is adjusted so that zero counts are excluded, but the probabilities in the function still sum to 1.0, as is required for a probability function.

Having data with an excessive number of zero counts is another problem for many count models. For example, given a specific mean value for a Poisson distribution of counts, the probability of having zero counts is defined by the PDF. When the mean is 2 or 3, the probability of having zero counts is quite good. But for a mean of 10, for instance, the Poisson PDF specifies that the probability of a zero count is very near zero. If your data have a mean value of 5 and some 30% of the count observations consist of zeros, there is a problem. The expected percentage of zero counts on the basis of the Poisson PDF is well under 1%. An adjustment must be made. Typically, analysts use either a two-part hurdle model or a mixture model, such as zero-inflated Poisson (ZIP) or zero-inflated negative binomial (ZINB). We will also use a zero-inflated PIG (ZIPIG), as well as other models.

Hurdle models are nearly always constructed as a two-part 0,1 response logistic or probit regression and a zero-truncated count model. The logistic component models the probability of obtaining a nonzero count. After separating the data into two components, the software creates a binary variable where all counts greater than zero (0) are assigned the value of one (1). Zeros in the count model are zeros in the logit component. The count component truncates or drops observations with zero values for the original counts and, for example, models the data as a zero-truncated Poisson. The model I describe here is a Poisson-logit hurdle model. Be aware, though, that the binary component of the hurdle model need not itself be from a binary model. An analyst can use a censored Poisson model, with right censoring at 1, and employ it as the binary component. This may not make much sense at this point, but when we address censored models later in the book, it will be clear. Most hurdle models use a logistic or probit regression for the binary component of the hurdle.

Zero-inflated models are not simply two-part models that can be estimated separately, as are hurdle models. Rather, zero-inflated models are mixture models. They use logistic or probit regression for the binary component, but both components – the binary and count – include the same zero counts when being estimated. The overlap of zero counts means that the mixture of Bernoulli (the distribution used in binary logistic regression) and Poisson distributions must be adjusted so that the resulting PDF sums to one. The statistical software takes care of this for you, but this overlap is important to remember since some statisticians – especially in ecology – interpret the binary zeros differently from the count zeros. I discuss this in more detail in Chapter 7. It should also be understood that zero-inflated models, unlike hurdle models, structure the binary component so that it models zeros instead of ones.

The count models described here are the only ones known by many analysts. However, several three-parameter count models have been developed that can be used on count data that fail to fit any of the standard count probability distributions, including mixtures of distributions. We will focus our attention on one of these models – the generalized NBP negative binomial model. The NBP model parameterizes the exponent on the second term of the negative binomial variance. Recall that the negative binomial variance function is $\mu + \alpha\mu^2$. We may symbolize the parameter as ρ (rho), representing the power – $\mu + \alpha\mu^\rho$. μ, α, and ρ are all parameters to be estimated.

Actually, there is another model we have only alluded to in our discussion thus far, the so-called NB1 model. It has a variance function of $\mu + \alpha\mu^1$ or simply $\mu + \alpha\mu$. Since it has a linear form, it is called a linear negative binomial; the traditional negative binomial is sometimes referred to as the quadratic negative binomial because of the square exponent. In any case, foremost use of the NBP model is to have it determine whether the data prefer NB1 or NB2. After estimating the NBP model, if ρ is close to 2, the analyst should use NB2 rather than NB1. I suggest letting the model speak for itself. If *alpha* is 0.5 and ρ is 1.8, then that should be the reported model. The NBP will be discussed in Chapter 5.

There are several other types of count models that are important in the analyst's statistical toolbox. At times you will have data that have been truncated or censored. We have already mentioned zero-truncated models. However, it is possible that the data theoretically come from population data that cannot have count values below 3, or perhaps above 10, or even to either side of two counts (e.g., 7 to 10). If values are truncated at the low end of the counts, the model is said to be left truncated; if they cannot exist higher than some cut point, the model is right truncated. Interval truncation exists when counts only exist between specific count values.

Censoring occurs where counts can possibly exist, but because of the study design or other factors they are not observed for these data. Left, right, and interval censoring can occur.

Finite mixture models are seldom used but can be exactly what you are looking for in a model. What if counts from more than one source are occurring? In the probabilistic sense, what if counts are coming into the data you are modeling from more than one data-generating mechanism? Perhaps 25% of the counts in your data are distributed as Poisson with a specific mean and the rest as Poisson with another mean value. A finite mixture model can ferret out the percentage distributions inherent in count data and provide means and coefficients/standard errors for each distributional component in the data. This is currently an underutilized class of models that can allow an analyst to have a much better understanding of various types of data situations than if they were modeled using standard Poisson or negative binomial methods.

I mentioned at the outset that the majority of count models to be discussed in this book are based on probability distributions. Nonparametric models, which include models with smoothers applied to specific continuous data in your model, can assist in fitting a model. Generalized additive models (GAMs) are used to assess the linearity of continuous predictors with respect to the response and provide information concerning what type of transform is needed to effect linearity. GAMs are not usually employed as a model to be used to predict or classify data.

Quantile count models are also nonparametric models but are used to describe the empirical distribution underlying one's data. Quantile count models are used when a parametric distribution, or mixture of distributions, cannot be identified. This is a new class of count model, which may well enjoy more popularity in the future. I discuss GAMs and quantile models briefly in Chapter 9. See Hilbe (2011) for more on the subject. A thorough analysis of GAMs for this class of models is given by Zuur (2012).

Finally, what happens if you have only a small amount of data or if your data are highly unbalanced? For example, a five-observation data set is almost always too small to appropriately model using standard asymptotic methods. Or consider a ten-observation data set for which a single binary predictor consists of nine 1's (90%) and one 0 (10%). The data are unbalanced. Or suppose that we have a three-level categorical predictor in a model with 20 observations, where level 1 has 10 observations and level 2 has 8 observations, leaving level 3 with only 2 observations. This is also an example of unbalanced data. For such models, using standard estimation methods will very likely fail to result in convergence, and if the model does converge, one or more standard errors will very likely be inflated. Exact Poisson regression should be attempted in such a situation.

Bayesian modeling is considered in the final chapter of the book. It is appropriate when you wish to have constraints on a predictor or to provide information about a predictor or predictors in a model in addition to the information already available given the predictor. It is also useful when there does not appear to be a PDF underlying the data to be modeled. Using a Markov Chain Monte Carlo (MCMC) sampling algorithm, a well-fitted empirical distribution can usually be found for which the user can obtain a mean and standard deviation and 95% quantiles. These translate to a predictor coefficient, standard error, and what is termed a *credible interval*.

There are other count models discussed in the literature, to be sure, but they are not commonly used by analysts for research. On the other hand, I have added several newer models to this book that I believe will become frequently used in research because of the current availability of software support and on account of their favorable properties. Some specialized models not addressed in this book will be reviewed on the book's web site.

1.4 ESTIMATION – THE MODELING PROCESS

1.4.1 Software for Modeling

All but a very few of the count models discussed in this book are estimated using two general methods: IRLS and full maximum likelihood estimation

(MLE). IRLS is an acronym meaning "iteratively reweighted least squares," which is the traditional method used to estimate statistical models from the class of *generalized linear models* (GLMs). IRLS is in fact based on a simplification of maximum likelihood that can occur when the models to be estimated are members of the one-parameter exponential family of probability distributions. This includes Poisson and negative binomial regressions, where the negative binomial dispersion parameter is entered into the GLM algorithm as a constant, as mentioned earlier in this chapter.

There are some situations where the parameters of a model cannot be estimated using either IRLS or full maximum likelihood estimation, such as mixed-effects models. Most applications use some variety of quadrature to estimate mixed-effects models, although a number of analysts are beginning to employ Bayesian modeling techniques (Zuur, Hilbe, and Ieno 2013). Mixed-effects models generally structure the data to be modeled in panels. When modeling longitudinal data, for example, panels contain the observations made over time for each individual in the model. If patients are being tested each year for five years, the data will have five observations for each individual – a panel. Since probability functions assume that each element in the distribution is independent of the others, when data are structured in panels it is clear that this assumption is violated. A number of models in addition to mixed models exist for dealing with longitudinal panel models; for example, generalized estimating equations (GEEs), which are estimated using a variety of IRLS algorithm.

Bayesian models, discussed in the final chapter, use a sampling algorithm referred to as Markov Chain Monte Carlo (MCMC) to develop a posterior distribution of the data. There are a wide variety of MCMC type algorithms, the two foremost being Metropolis–Hastings and Gibbs sampling. There are a number of variations of these algorithms as well. We discuss this in more detail in Section 9.8.

1.4.2 Maximum Likelihood Estimation[4]

Recall that we have characterized the goal of modeling to be the identification of a probability distribution, or mixture of distributions, and estimation of

---

[4] This section and Section 1.4.3 are optional and require a knowledge of basic calculus. They are included for those who wish to obtain an overview of count model estimation, but they are not required for reading the subsequent chapters. I do suggest reading the first part of this section, though, to learn about calculating Poisson probabilities.

its unknown parameter or parameters, which can be said to have generated the population of data from which we are modeling a random sample. This may not be how we actually think about the (frequency-based) modeling process, but it is nevertheless its logic. A probability distribution is itself defined in terms of the values of its parameter or parameters.

We may define a probability distribution function for count models as

$$f(y \mid \theta, \phi) \quad \text{or} \quad f(y; \theta, \phi) \tag{1.10}$$

where y is the count response, θ is the canonical location parameter or link function, and ϕ is the scale parameter. Count models such as Poisson regression set the scale to a value of one; other more complex count models have a scale parameter, and in some cases more than one. The outcome y, of course, has the distributional properties appropriate to the statistical model used in estimation. Probability functions determine properties of the response, as adjusted by explanatory predictors, and given values of the mean and scale parameters.

A probability function generates or describes data on the basis of parameters. When we are actually modeling data, though, we know what the data are. They have been collected for our study, or we use data that have been previously published. In modeling, we seek the value of the probability distribution, as defined by specific unknown parameters, that makes the data we have most likely. In order to do this, we in effect invert the relationship of y and the PDF parameters, creating what is called a *likelihood function*. The likelihood function can be defined as

$$L(\theta, \phi \mid y) \quad \text{or} \quad L(\theta, \phi; y) \tag{1.11}$$

We can clarify the difference between the probability and likelihood functions as follows:

A probability function generates data on the basis of known parameters.

A likelihood function determines parameter values on the basis of known data.

Given that we do know the data we are modeling, and given the fact that we are attempting to estimate the parameters of the probability function that (theoretically) generate the data we are modeling, we need to employ a likelihood function in order to determine the parameters that make the model data most likely to be the case. Expressed in another way, when modeling we are asking what parameter values of a given PDF most likely generated the data we have to model.

To make these concepts better understood, recall that the Poisson probability distribution has a single parameter and the negative binomial has two. Each of these parameters can theoretically have an infinite number of parameter values. Once we select a Poisson probability, for example, to describe our data, we need to refine the description by determining the location parameter. We do this by maximizing, or optimizing, the Poisson likelihood function. The method we use to maximize the likelihood function in question is referred to as *maximum likelihood estimation.*

Because of numerical considerations, statisticians maximize the log of the likelihood rather than the likelihood function itself. We will find throughout this book that the log-likelihood is one of the most well-used statistics in modeling. Maximum likelihood estimation (MLE) is in fact maximum log-likelihood estimation, although we rarely hear the term used in this manner.

Maximization of the log-likelihood function involves taking the partial derivatives of the function, setting the resulting equation to 0, and solving for parameter values. The first derivative of the log-likelihood function, with respect to the coefficients, is called the score function, U. The second derivative is a matrix called the Hessian matrix. The standard errors of the predictors in the model are obtained by taking the square root of the diagonal terms of the negative inverse Hessian, $-H^{-1}$. I refer you to Hilbe (2011) and Hilbe and Robinson (2013), which provide an in-depth evaluation of both maximum likelihood and IRLS methodology.

Before leaving this overview of maximum likelihood estimation, I'll give an example of how this works, beginning with the Poisson probability distribution. It is the basic count model. If you understand the logic of estimation, it will be easier to understand when and why things can go wrong in modeling.

In its simplest form, the Poisson probability distribution can be expressed as

$$f(y; \mu) = \frac{e^{-\mu_i} \mu_i^{y_i}}{y_i!} \qquad (1.12)$$

where y represents a variable consisting of count values and μ is the expected or predicted mean of the count variable y. $y!$, meaning y-factorial, is the product of counts up to a specific count value, y. $f(y; \mu)$ indicates the probability of y given or based on the value of the mean. The subscripts indicate that the distribution describes each observation in the data. Technically, equation (1.12) should have a product sign before the terms on the right side of the equation, indicating that the function is the joint probability of the observations in the model being described by the function. The joint probability can be represented as Pr(*Obs1 and Obs2 and Obs3* $\cdots$ *and* $\cdots$ *Obs-n*). If the observations are independent of one another, the joint probability may be expressed as the product of the function across observations. Since we assume that the observations being modeled using a Poisson probability distribution are independent, the product sign is assumed but generally not displayed.

For an example of a simple Poisson model, suppose that we count the number of customers waiting at the entrance for the bank to open. We do this each morning for a week, including Saturdays. A more ambitious study might be to count waiting customers for each day over a month, or even for a six-month period. However, we'll evaluate the numbers for a single week. To be specific, we count the number of customers waiting for service at the time the bank opens:

```
Mon =   4;  Tue =   2; Wed =    0; Thu =    3; Fri =   1; Sat =   2
```

The mean number of counts over the six days is 2. If a Poisson probability distribution with a mean, μ, of 2 truly describes this list or vector of counts, we can calculate the probability of no (0) customers being in line, or any other number of customers, using the Poisson probability function:

```
CUSTOMERS
Prob(0) :       . di  exp(-2) * (2^0)/0 = .13533528

. di exp(-2)* (2^0)/exp(lnfactorial(0))
.13533528
```

or simply

```
. di exp(-2)
.13533528
```

To calculate the probability of one person being in line and then two, three, and four, we have

```
1 . di exp(-2)* (2^1)/exp(lnfactorial(1)) = .27067057
2 . di exp(-2)* (2^2)/exp(lnfactorial(2)) = .27067057
3 . di exp(-2)* (2^3)/exp(lnfactorial(3)) = .18044704
4 . di exp(-2)* (2^4)/exp(lnfactorial(4)) = .09022352
```

TABLE 1.4. R Poisson Probabilities for y from 0 through 4

```
> y <- c(4, 2, 0,3, 1, 2)
> y0 <- exp(-2)* (2^0)/factorial(0)
> y1 <- exp(-2)* (2^1)/factorial(1)
> y2 <- exp(-2)* (2^2)/factorial(2)
> y3 <- exp(-2)* (2^3)/factorial(3)
> y4 <- exp(-2)* (2^4)/factorial(4)
> poisProb <- c(y0, y1, y2, y3, y4); poisProb
[1] 0.13533528 0.27067057 0.27067057 0.18044704 0.09022352
# OR
> dpois(0:4, lambda=2)
[1] 0.13533528 0.27067057 0.27067057 0.18044704 0.09022352
# CUMULATIVE
> ppois(0:4, lambda=2)
[1] 0.1353353 0.4060058 0.6766764 0.8571235 0.9473470
# to plot a histogram
py <- 0:4
plot(poisProb ~ py, xlim=c(0,4), type="o", main="Poisson Prob 0-4:
Mean=2")
```

Note: Stata's **poisson(mu,y)** function provides a cumulative Poisson probability for a given mean and count term, y. **poissonp(mu, y)** gives a specific probability for a mean and y value; for example, for a mean and count (y) of 2

```
. di poissonp(2,2)
.27067057
```

the cumulative values of y from 0 through 2,

```
. di poisson(2,2)
.67667642
```

In place of a constant value for the mean, *mu*, let's vary it, giving it four different values:

$$\mu = mu = \{0.5, 1, 3, 5\}$$

Plotting the values allows us to observe the differences in the shapes of the distributions. This type of graph will prove to be of use later when we discuss

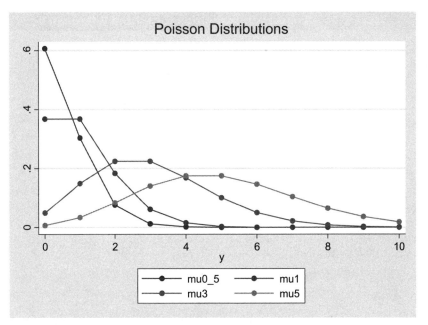

FIGURE 1.3. Poisson distributions.

zero-truncated and zero-inflated models. The following code can be pasted into the Stata *doeditor* and run, producing the graphic in Figure 1.3. The R code for this figure is given in Table 1.5.

```
STATA CODE
clear
set obs 11
gen y = _n-1
gen mu = .
gen mu0_5 = (exp(-.5)* .5^y)/exp(lngamma(y+1))
forvalues i = 1(2)6 {
    gen mu'i' = (exp(-'i')*'i'^y)/exp(lngamma(y+1))
    }
graph twoway connected mu0_5 mu1 mu3 mu5 y, title("Poisson
Distributions")
```

Note that mean values under 1 are shaped like negative exponential distributions. The greater the mean, the more normal the shape of its appearance.

When we wish to adjust the values of the probability of customers waiting by such factors as "Is it raining or snowing?", "Is it the first of the month – directly after customers may have been paid?", "Is there an epidemic in the

TABLE 1.5. R Code for Figure 1.3

```
m<- c(0.5,1,3,5)              #Poisson means
y<- 0:11                      #Observed counts
layout(1)
for (i in 1:length(m)) {
  p<- dpois(y, m[i])          #poisson pdf
  if (i==1) {
  plot(y, p, col=i, type='l', lty=i)
  } else {
  lines(y, p, col=i, lty=i)
  }
}
```

area?", and so forth, it is necessary to construct a model based on discovering what parameter value makes it most likely that this specific list of count values occurred. This means that we reparameterize the Poisson PDF and log both sides to derive the log-likelihood function.

The equation for the Poisson log-likelihood, showing summation across the observations as discussed earlier, can be expressed as

$$\mathcal{L}(\mu; y) = \sum_{i=1}^{n} \{y_i \ln(\mu_i) - \mu_i - \ln(y_i!)\} \tag{1.13}$$

Recall that the linear predictor of the Poisson model is xb, or $x\beta$. Recall also that $x\beta = \ln(\mu)$. This entails that $\mu = \exp(x\beta)$. $\exp(x\beta)$ is called the *inverse link function*, which defines μ. $\exp(x\beta)$ also defines μ for the negative binomial model, as well as for most models based on the Poisson and negative binomial that are used in research. Because of this relationship, $\mu = \exp(x\beta)$ is also referred to as the *exponential mean function*. If we substitute $\exp(x\beta)$ for μ (1.14) may be expressed as

$$\mathcal{L}(\beta; y) = \sum_{i=1}^{n} \{y_i(x_i'\beta) - \exp(x_i'\beta) - \ln \Gamma(y_i + 1)\} \tag{1.14}$$

The first derivative of the preceding log-likelihood function with respect to the coefficients (β), which are also called the parameters when modeling, provides the *gradient*, or *score*, of the Poisson log-likelihood:

$$\frac{\partial(\mathcal{L}(\beta; y))}{\partial \beta} = \sum_{i=1}^{n} (y_i - \exp(x_i'\beta))x_i' \tag{1.15}$$

TABLE 1.6. Newton–Raphson Maximum Likelihood Estimation

```
Initialize β       # Provide initial or starting values for estimates
WHILE (ABS(βn-βo)>tol & ABS(Δn-Δo)>tol) {
    g   = ∂L/∂β    # gradient: 1st derivative of log-likelihood wrt β
    H   = ∂²L/∂β²  # Hessian: 2nd derivative of log-likelihood wrt β
    βo  = βn
    βn  = βo - H⁻¹g # updated maximum likelihood estimates
    Lo  = Ln
    Ln                   # new log-likelihood value
}
```

Setting equation (1.15) to zero (0) provides for the solution of the parameter estimates.

The Hessian matrix is calculated as the second derivative of the log-likelihood function and is negative definite for β. For the Poisson, it may be expressed as

$$\frac{\partial^2(\mathcal{L}(\beta; y))}{\partial\beta\partial\beta'} = -\sum_{i=1}^{n}(\exp(x_i'\beta))x_i x_i' \qquad (1.16)$$

Estimation of the maximum likelihood variance–covariance matrix is based on the negative inverse of the Hessian, often represented as Σ (but not to be understood as a sum symbol), given as

$$\Sigma = -H^{-1} = \left[\sum_{i=1}^{n}(\exp(x_i'\beta))x_i x_j'\right]^{-1} \qquad (1.17)$$

The square roots of the respective terms on the diagonal of the negative inverse Hessian are the values of predictor standard errors. A Newton–Raphson type algorithm (see Table 1.6) can be used for the maximum likelihood estimation of the parameters

$$\beta_{r+1} = \beta_r - H^{-1}g \qquad (1.18)$$

which is the standard form of the maximum likelihood estimating equation.

The tools have now been developed to construct a full maximum likelihood Poisson algorithm. The algorithm typically updates estimates based on the value of the log-likelihood function. When the difference between old and updated values is less than a specified tolerance level – usually 10^{-6} – iteration stops and the values of the various statistics are at their maximum likelihood estimated values. Other iteration criteria have been employed, but all generally result in the same estimates. The scheme for maximum likelihood estimation follows. It is the basis for models being estimated using this method. βo indicates estimation at the old value of the parameters (coefficients); βn is the new, updated value. Likewise for the likelihood, *L. tol* indicates the level of tolerance needed for convergence, which is typically set at 10^{-6}.

Refer to Section 2.4 for examples of both Stata and R maximum likelihood code. Example code is provided for a Poisson regression estimated using maximum likelihood. Real data are used for the no-frills Stata and R programs, but model coefficients and associated statistical values are calculated and displayed.

1.4.3 Generalized Linear Models and IRLS Estimation

Although the vast majority of models we discuss in this book are modeled using the full maximum likelihood methods of estimation described in the previous section, the two foremost count models – Poisson and negative binomial regression – are often estimated as a generalized linear model. The **glm** and **glm.nb** functions are the default method for estimating Poisson and negative binomial models using R. Stata users may use the commands **poisson** and **nbreg** for maximum likelihood estimation, or **glm** using an IRLS algorithm. The SAS **Genmod** procedure for generalized linear models and several other GLM extensions is a maximum likelihood procedure cast in an IRLS format. The Poisson and negative binomial models for SPSS are available only in the **Genlin** procedure, a GLM algorithm. Limdep, an econometric software package with extensive count-modeling capabilities, does not use IRLS but rather MLE, quadrature, and simulation.

The Poisson distribution is a single-parameter member of the exponential family of probability functions. The exponential family of distributions is symbolized in a variety of ways. A standard formulation for one-parameter models is expressed as

$$f(y; \theta) = \exp\{y_i \theta_i - b(\theta_i) + c(y_i)\} \qquad (1.19)$$

θ_i is the canonical parameter or link function, defined as ln(mu) for the Poisson and NB models. $b(\theta_i)$ is the cumulant. The first and second derivatives of $b(\theta_i)$ define the mean and variance. $c(y)$ is the normalization term, as given in equation (1.10).

TABLE 1.7. The IRLS Fitting Algorithm

1. Initialize the expected response, μ, and the linear predictor, η, or $g(\mu)$.
2. Compute the weights as $W^{-1} = Vg'(\mu)^2$, where $g'(\mu)$ is the derivative of the link function and V is the variance, defined as the second derivative of the cumulant, $b''(\theta)$.
3. Compute a working response, a one-term Taylor linearization of the log-likelihood function, with a standard form of (using no subscripts)
   $z = \eta + (y - \mu)g'(\mu)$.
4. Regress z on predictors $X_1 \ldots X_n$ with weights W to obtain updates on the vector of parameter estimates, β.
5. Compute η, or $X\beta$, based on the regression estimates.
6. Compute μ, or $E(y)$, as $g^{-1}(\mu)$.
7. Compute the deviance or log-likelihood function.
8. Iterate until the change in deviance or log-likelihood between two iterations is below a specified level of tolerance, or threshold.

Source: Hilbe (2007a).

Derivation of the iteratively reweighted least squares (IRLS) algorithm is based on a modification of a two-term Taylor expansion of the log-likelihood function where for count models y is a vector of count values and θ is the parameter or parameters of the probability function generating y. The logic of the IRLS estimating algorithm is given in Table 1.7, where

$g(\mu)$ is the link function: for Poisson and NB, $\ln(mu)$

$g'(\mu)$ is the derivative of the link, or $1/mu$

$g^{-1}(\mu)$ is the inverse link function, $\exp(xb)$

$b'(\theta)$ is mu, the mean

V is $b''(\theta)$, the variance: Poisson: mu; NB: $mu + \alpha^* mu^2$; PIG: $mu + \alpha^* mu^3$

Deviance: Will be discussed in Chapter 3. See equations (3.1) and (3.2) and related discussion. It is similar to a likelihood ratio test of the saturated log-likelihood minus the model log-likelihood: $D = 2\{LLs - LLm\}$.

The GLM IRLS algorithm for the general case is presented in Table 1.7. The algorithm can be used for any member of the GLM family. The Poisson model will be the subject of Chapter 2. I will provide example code for Poisson regression using an IRLS algorithm. Both Stata and R examples are provided. SAS code is in the Appendix.

1.5 SUMMARY

There are several important points to take from this chapter. The first point we have made is that most analysts are aware that Poisson and negative binomial regressions are used to model count response data, but they have little knowledge of other count models or under what conditions they may be desirable. They may have heard of overdispersion but not be fully aware of what it is or how it relates to the Poisson model. We have emphasized, however, that overdispersion is a very real problem for Poisson regression and occurs when the variance of the Poisson model is greater than the mean. Overdispersion also occurs more generally when there is more variability in the model data than is assumed on the basis of the probability distribution on which the model is based. Negative binomial regression, having a second parameter, is now a standard way of modeling overdispersed Poisson data.

Negative binomial regression may not adequately adjust for the amount of overdispersion in a Poisson model, depending on what gives rise to the overdispersion or underdispersion. A number of other models based on the Poisson and negative binomial models have been designed to appropriately compensate for overdispersion.

Another important point to take from this opening chapter is that when we are modeling data we are in fact modeling an underlying probability distribution (PDF) or mixture of distributions. When modeling, the analyst selects a PDF or mixture of PDFs that is believed to best describe the data to be modeled. Once a selection is made, the analyst attempts to estimate the parameters that make the data most likely to be the case. Analysts generally do this using maximum likelihood estimation. IRLS estimation, which is the algorithm typically used in generalized linear model (GLM) software, is a variety of maximum likelihood. It is simplified because GLM models are members of the single-parameter exponential family of distributions.

The data that are being modeled using maximum likelihood techniques are also thought of as a random sample from a greater population of data, so the model is in fact of a distribution that theoretically generates the population data from which the model data are a random sample. Table 1.8 provides a rough summary of these relationships. A summary of the relationship of modeling to probability can be expressed as

> *The foremost goal of modeling is the identification of a probability distribution, or mixture of distributions, and estimation of its unknown*

TABLE 1.8. The Modeling Process
1. Describe the response data. Be confident that the data being modeled are theoretically a random sample from a greater population of data.
2. Select a probability distribution, or mixture of distributions, that you believe best describes the data being modeled.
3. Model: Calculate unbiased estimates of the parameter(s) of the PDF that makes your model data most likely. Include all relevant predictors.
4. Calculate the fitted or predicted values of the response based on the model.
5. Assess the differences between the estimated or fitted values and actual values; i.e., test the closeness of y and μ values across all observations.
6. Evaluate fit test statistics; compare them with other models on the same data.

parameter or parameters, that can be said to have generated the population of data from which we are modeling a random sample.

This is not normally how we think of the modeling process, but based on the predominant frequency-based interpretation of modeling, it makes sense. At its core, it is consistent with the view that when modeling we are attempting to discover the unbiased parameter values of a probability distribution, or mixture of distributions, that make the data we are modeling most likely to be the case. In the following chapter, we discuss in more detail how we make this discovery.

## CHAPTER 2

Poisson Regression

SOME POINTS OF DISCUSSION

- How do Poisson models differ from traditional linear regression models?
- What are the distributional assumptions of the Poisson regression model? For any count model?
- What is the dispersion statistic? How is it calculated?
- What is the relationship of Poisson standard errors to the dispersion statistic?
- What is apparent overdisperson? How do we deal with it?
- How can a synthetic Monte Carlo Poisson model be developed?
- How are Poisson coefficients and rate-parameterized coefficients interpreted?
- What are marginal effects, partial effects, and discrete change with respect to count models?

Poisson regression is fundamental to the modeling of count data. It was the first model specifically used to model counts, and it still stands at the base of the many types of count models available to analysts. However, it was emphasized in the last chapter that because of the Poisson distributional assumption of equidispersion, using a Poisson model on real study data is usually unsatisfactory. It is sometimes possible to make adjustments to the Poisson model that remedy the problem of under- or overdispersion, but unfortunately often this is not possible. In this chapter, which is central to the book, we look at the nature of Poisson regression and provide guidelines on how to construct, interpret, and evaluate Poisson models as to their fit. The majority of fit tests we use for a Poisson model will be applicable to the more advanced count models discussed later. This will also be true of how we interpret count model coefficients and how predictions are produced.

We begin by providing a list of assumptions that Poisson models must abide by in order to appropriately fit the data. This is followed by guidelines on how an analyst can test to determine whether any of these assumptions are violated in the data being modeled. The listing will be in summary form. Typically such a summary is at the end of a chapter, but I am placing it at the beginning to serve as a guideline for you to use when modeling and as a notice of subsequent discussion in the chapter.

2.1 POISSON MODEL ASSUMPTIONS

When analysts have count data to model, they must first check the data to see whether there are any major violations of the assumptions on which the basic Poisson model is based (see Table 2.1). Poisson is the standard count model, and the Poisson probability distribution function (PDF) is the assumed manner in which we expect count data to be structured unless we have reason to suspect otherwise. All other count models are adjustments or variations from the basic Poisson model.

Each of these assumptions should be tested.

1. The response or dependent variable of a Poisson model must be a count. If we wish to model a continuous variable, another model must be selected.
2. This criterion is often violated when a Poisson model is employed on a count response term that does not include the possibility of having zero counts (i.e., zero counts must be allowed as a possibility even though there may not be any in a given modeling situation). For example, when dealing with hospital outcomes data, patient length of stay (LOS) is often modeled. But LOS values exclude zero days. As soon as a patient is admitted, LOS = 1. That is, the range of LOS data is $Y \geq 1$. A count term in a model that could include zero counts but happens not to does not violate this criterion.

Each possible count in a Poisson distribution with a mean value of μ has a specific probability of occurrence. The sum of these probabilities must be

---

TABLE 2.1. Basic Poisson Assumptions

1. The distribution is discrete with a single parameter, the mean, which is usually symbolized as either λ (*lambda*) or μ (*mu*). The mean is also understood as a rate parameter. It is the expected number of times that an item or event occurs per unit of time, area, or volume.
2. The response terms, or y values, are nonnegative integers; i.e., the distribution allows for the possibility of counts where $Y \geq 0$.
3. Observations are independent of one another.
4. No cell of observed counts has substantially more or less than what is expected based on the mean of the empirical distribution. For example, the data should not have more zero counts than is expected based on a Poisson distribution with a given mean. As the value of μ increases, the probability of zero (0) counts is reduced.
5. The mean and variance of the model are identical, or at least nearly the same; i.e., Poisson distributions with higher mean values have correspondingly greater variability.
6. The Pearson Chi2 dispersion statistic has a value approximating 1.0. A value of 1.0 results when the observed and predicted variances of the response are the same.

---

one (1). However, if zero is excluded as a possibility, then the sum of probabilities will always be less than 1. In order to properly model the count data having a low mean value, a zero-truncated Poisson distribution must be used for modeling. We discuss this in more detail in Chapter 7.

3. Two ways in which independence may be tested are:
   a. Check to determine whether the data are structured in panels (i.e., check whether the data are clustered or are a collection of longitudinal data). If they are, then they are not independent. When we model otherwise panel data as though they are independent, we say that the data have been "pooled" (i.e., we assume that the panel effect is not enough to violate the independence criterion).
   b. Check the difference between the model SEs and the SEs adjusted by
      i. employing a robust sandwich estimator to the SEs, or
      ii. bootstrapping the SEs
         or
      iii. checking SEs scaled by the dispersion statistic (i.e., model SEs multiplied by the square root of the Pearson Chi2 dispersion).

If the SEs differ considerably, then the data are correlated. I suggest applying all three adjustments to the model SEs. If all differ from the model SEs, it is highly likely that the independence criterion has been violated.

4. This criterion may be tested by calculating the percentage of zeros in the empirical distribution (i.e., the distribution of the count response variable). Compare that with the frequency of zero counts expected or predicted based on a Poisson PDF with the mean determined for the observed distribution. If the compared frequencies substantially differ, then a violation exists in the distributional assumption. An example was given in Section 1.4.2, where we calculated the probability of six observations with count values of 0 through 4. The mean of the six counts was calculated to be 2. The probability of a 0 count for a Poisson distribution with a mean of 2 is .135, or 13.5%. If we have a count variable with a mean of 2 but 25% of the counts are 0, this criterion has been violated. It is generally permissible to have small variations in the differences between actual and expected counts, but it is usually wise to check using a Chi2 test, which is described later in this chapter. The relationship of the probability of zero counts and the value of μ will be explored in greater detail in Section 7.1, where we discuss truncated zeros.

5. Calculate the mean and variance of the empirical (observed) count response. Determine whether the two values differ.

6. The Pearson Chi2 test is performed as follows:
   a. Tabulate the observed variance of the count variable being modeled. Compare it with the predicted variance. If the two values differ, the model is extradispersed. If the observed variance is greater than the expected or predicted variance, the model is "Poisson overdispersed"; if it is less, the model is Poisson underdispersed. Most count data are in fact Poisson overdispersed.
   b. Estimate the desired full Poisson model, checking the value of the Pearson Chi2 dispersion. If the value of the dispersion varies from 1.0, the model is "Poisson extradispersed." A boundary *likelihood ratio test* can be used to assess overdispersion. A generalized Poisson model can test for the statistical significance of either under- or overdispersion.

When the preceding criteria are violated, the model generally appears as Poisson extradispersed and in particular as Poisson overdispersed. As such, the criteria are all related to some degree, some more than others. An indication of extradispersion, whether under- or over-, is given by the dispersion statistic, as described in part 6b of the list.

If a model under consideration fails to violate any of the Poisson distributional assumptions, then we can model the data using a standard Poisson model. If the data show evidence of extradispersion, we either must

1. employ an alternative count model that adjusts for the type of distributional assumption violated in the data, or
2. if no clear distributional violation is apparent (e.g., no possibility of zero counts, or excessive zero counts) and the dispersion still differs from one (1), the model may in fact be only apparently extradispersed, as discussed later in the text. The researcher should check each of the alternative "reasons" for extradispersion.

In Table 2.2, we summarize the steps to consider if a Poisson model appears overdispersed.

2.2 Apparent Overdispersion

It was mentioned in the previous section that a Poisson model may appear to be overdispersed when the overdispersion is only apparent, not real. This is a very important point to remember when modeling count data. Too often an analyst will construct a complex negative binomial model only to later discover that if they had perhaps made a simple amendment to the Poisson model the overdispersion would have disappeared. Table 2.3 lists criteria for apparent Poisson overdispersion.

If we are confident that Poisson model assumptions have been violated, even after adjustments for possible apparent extradispersion, then we may conclude that a given data situation is Poisson extradispersed. Note that data may be Poisson overdispersed yet negative binomial equidispersed, or some other variation. A model is overdispersed, for example, with respect to a given distribution, not necessarily all distributions.

When a model is judged to be extradispersed, we first determine whether it is under- or overdispersed. For a Poisson model, it is overdispersed, even after adjustments for apparent overdispersion have been made, if the dispersion statistic is greater than 1 and is underdispersed if the dispersion is less than 1.

TABLE 2.2. First Steps to Take If the Poisson Model Appears Overdispersed

Amend the model standard errors by (see Section 3.4 for details)

- *multplying them by the square root of the Poisson dispersion statistic, defined as $(y - \mu)^2/\mu$;*
- *using a robust or sandwich variance estimator. The resulting standard errors are also called robust or empirical standard errors. The model using robust standard errors is at times referred to as a quasi-maximum likelihood model or as a quasi-likelihood model;*
- *bootstrapping.*

Employ an alternative count model that adjusts for the type of distributional assumption violated in the data; e.g., use a:

- *zero-truncated Poisson (ZTP) model if there is no possibility of 0 counts in the data;*
- *zero-inflated Poisson (ZIP) model if there are more 0 counts than are expected based on the Poisson PDF for a given mean or 0 counts have a different source than positive counts. Is there a theory as to why there are too many 0's?*
- *hurdle model if there are more, or fewer, 0 counts than are expected based on the Poisson PDF for a given mean or 0 counts have a different source than positive counts.*

If 0 counts appear to have a different source than the positive counts, use:

- *a truncated Poisson model if count values are not possible within the range of counts;*
- *a censored Poisson if count values are not recorded for some reason within the range of counts;*
- *negative binomial versions of the preceding "solutions."*

If overdispersion is evident from the dispersion statistic, but we have no clear idea what is causing the overdispersion:

- *check to determine if the model is only apparently overdispersed (discussed next);*
- *use a heterogeneous negative binomial to determine which predictors contribute to the Poisson overdispersion (to be discussed later);*
- *use a negative binomial (NB2) model, checking its dispersion statistic to determine if it under- or overadjusts for the extra dispersion in the Poisson model (to be discussed: what happens if the NB model under- or overadjusts?);*
- *use a Poisson inverse Gaussian (PIG) model (or alternative count model; e.g., generalized Poisson), and compare the AIC and BIC tests with NB2. Use a model with (substantially) lower AIC and BIC statistics. If there is no statistical difference, use NB2 and then test (to be discussed).*

TABLE 2.3. Criteria for Apparent Poisson Overdispersion

1. The model omits important explanatory predictors.
2. The data include outliers.
3. The model fails to include a sufficient number of interaction terms.
4. A predictor needs to be transformed to another scale.
5. There are situations where the data are too sparse and more data need to be collected and included in the model.
6. Missing values exist in the data but are not randomly distributed in the data; i.e., they are not missing at random (MAR).

The next stage in the process of modeling counts is to determine, if possible, the source of extradispersion. How one models largely depends on knowing why the model is extradispersed, or at least what is the most likely reason. We have already mentioned several reasons; for example, there may be far too many zero counts in the data given the mean of the distribution. In such a case, we may wish to attempt modeling the data using a group of models called zero-inflated mixture models. These models mix a count model with a separate model for the zero counts. This is typically a binomial logit model, but need not be. It must be remembered also that there may be more than one source of correlation in the data or more than one source of distributional violation. Care must be taken to account for all sources of difficulty and not only one. This can be a time-consuming procedure, but it is important if one is attempting to properly model a given data situation.

2.3 CONSTRUCTING A "TRUE" POISSON MODEL

I believe that it is important to see exactly what a "true" Poisson model looks like. We do this by creating a synthetic Poisson variable with specified parameter values. Poisson model software will then be able to model the data such that the model output reflects the Poisson variable we created. That is, the coefficients of the model will be nearly the same as the parameter values we gave the distribution. As a result, the dispersion statistic will have a value approximating 1.0. Modeling the data in this manner makes it abundantly clear how a statistical count model relates to a probability distribution that theoretically generates the population data from which the model samples but a random part.

For our example model, we generate a random Poisson count response variable adjusted by three explanatory predictors that are themselves generated by a uniform random number generator. We make each predictor into uniformly distributed continuous variables, assigning them a specific coefficient or slope. This sort of exercise is important for testing the assumptions of models, including model statistics.

We construct a data set with $x1$, $x2$, and $x3$ defined as random uniform variables. The intercept will be assigned a value of 1. The coefficients we give to the variables follow:

```
Intercept ==   1.00          X1          == 0.75
X2          == -1.25          X3          == 0.50
```

Stata code to create the simulated data consists of the following:

```
STATA CODE
. clear
. set obs 50000
. set seed 4590
. gen x1 = runiform()
. gen x2 = runiform()
. gen x3 = runiform()
. gen xb = 1 + 0.75*x1 - 1.25*x2 + .5*x3
. gen exb =  exp(xb)
. gen py = rpoisson(exb)
. tab py
```

py	Freq.	Percent	Cum.
0	4,579	9.16	9.16
1	9,081	18.16	27.32
2	10,231	20.46	47.78
3	8,687	17.37	65.16
4	6,481	12.96	78.12
.	.	.	.
14	16	0.03	99.98
15	7	0.01	99.99
16	3	0.01	100.00
Total	50,000	100.00	

The mean and median of the Poisson response are 3.0. The displayed output has been amended:

```
. sum py, detail
50%       3                     Mean     3.00764
. glm py x1 x2 x3, fam(poi) nolog        // non-numeric mid-header
                                         output deleted
Generalized linear models           No. of obs      =      50000
Optimization     : ML               Residual df     =      49996
                                    Scale parameter =          1
Deviance       =   54917.73016      (1/df) Deviance =   1.098442
Pearson        =   49942.18916      (1/df) Pearson  =   .9989237
                                    AIC             =   3.744693
Log likelihood =  -93613.32814      BIC             =  -486027.9
-----------------------------------------------------------------
             |              OIM
         py  |    Coef.  Std. Err.     z   P>|z|   [95% Conf. Interval]
-------------+---------------------------------------------------
         x1  |  .7502913  .0090475  82.93  0.000   .7325586    .768024
         x2  | -1.240165  .0092747 -133.71 0.000  -1.258343  -1.221987
         x3  |  .504346   .0089983  56.05  0.000   .4867096   .5219825
      _cons  |  .9957061    .00835 119.25  0.000   .9793404   1.012072
-----------------------------------------------------------------
. abic
AIC Statistic   =    3.744693       AIC*n       = 187234.66
BIC Statistic   =    3.744755       BIC(Stata)  = 187269.94
```

We set a seed value of 4590, which I just provided by chance. Using this same seed will guarantee that the same data and model is displayed. Dropping the seed will result in each run giving slightly different output. If we ran a Monte Carlo simulation of the data with perhaps 100 iterations, the parameter estimates would equal the values we assigned them, and the Pearson dispersion statistic, defined as the Pearson statistic divided by the model degrees of freedom, would equal 1.0. Note that the Pearson dispersion statistic in the preceding model is 0.9989237, with the parameter estimates approximating the values we specified. Note also that the deviance-based dispersion statistic is some 10% greater than 1. After hundreds of simulations, I have found that simulations of a true Poisson model always result in a Pearson dispersion statistic approximating 1. The deviance appears to vary between 1.03 and 1.13. I have also displayed a user command I wrote called **abic**, which provides two types each of AIC (Akaike Information Criterion) and

TABLE 2.4. R: Synthetic Poisson Model

```
library(MASS); library(COUNT); set.seed(4590); nobs <- 50000
x1 <- runif(nobs); x2 <- runif(nobs); x3 <- runif(nobs)
py <- rpois(nobs, exp(1 + 0.75*x1 - 1.25*x2 + .5*x3))
cnt <- table(py)
dataf <- data.frame(prop.table(table(py) ) )
dataf$cumulative <- cumsum(dataf$Freq)
datafall <- data.frame(cnt, dataf$Freq*100, dataf$cumulative * 100)
datafall; summary(py)
summary(py1 <- glm(py ~ x1 + x2 + x3, family=poisson))
confint.default(py1); py1$aic/(py1$df.null+1)
pr <- resid(py1, type = "pearson")
pchi2 <- sum(residuals(py1, type="pearson")^2)
disp <- pchi2/py1$df.residual; pchi2; disp
```

BIC (Bayesian Information Criterion) fit statistics. When comparing models as to fit, lower values of either the AIC or BIC indicate a better fit. We discuss both statistics in more depth later in the chapter.

Finally, it is clear that the coefficients we assigned to the random variates are closely approximated using the Poisson model. The coefficients are nearly identical to what we specified, the standard errors are tight, p-values are all 0.000, and the dispersion statistic is near 1.0. The model is a true Poisson model.

R code and output follow for the same example. I have given the R code (Table 2.4) the same apparent seed; realize, though, that the seed mechanisms differ between the two software packages. The results will be similar but not identical:

```
> datafall
```

	py	Freq	dataf.Freq...100	dataf.cumulative...100
1	0	4667	9.334	9.334
2	1	9134	18.268	27.602
3	2	10190	20.380	47.982
4	3	8814	17.628	65.610
.	.	.	.	.
15	14	15	0.030	99.980
16	15	8	0.016	99.996
17	16	2	0.004	100.000

```
> summary(py)
   Min. 1st Qu.  Median    Mean 3rd Qu.    Max.
  0.000   1.000   3.000   2.993   4.000  16.000
> ...
Coefficients:
             Estimate Std. Error z value Pr(>|z|)
(Intercept)  0.985628   0.008415  117.13   <2e-16 ***
x1           0.748436   0.009091   82.33   <2e-16 ***
x2          -1.243313   0.009298 -133.72   <2e-16 ***
x3           0.518207   0.009018   57.46   <2e-16 ***
---

    Null deviance: 84129  on 49999  degrees of freedom
Residual deviance: 55154  on 49996  degrees of freedom
AIC: 187059

> confint.default(py1)             # model confidence intervals
                 2.5 %      97.5 %
(Intercept)  0.9691352   1.0021199
x1           0.7306175   0.7662541
x2          -1.2615377  -1.2250893
x3           0.5005321   0.5358818
[1] 3.741183

#  load library msme and run "> P__disp(py1)"  to obtain
   Pearson Chi2 and dispersion
> pr <- resid(py1, type = "pearson")              # by-hand method
> pchi2 <- sum(residuals(py1, type="pearson")^2)
> disp <- pchi2/py1$df.residual
> pchi2; disp
[1] 49786.62
[1] 0.9958121
```

The model results are nearly the same, regardless of whether they are generated using Stata or R. Even with different seed values, the models are nearly identical. Note that the same seeds for Stata and R models result in slightly different values. The R dispersion statistic is 0.9958121.

To determine whether the single Poisson models we just ran are providing values that approximate the parameter values we assigned to the algorithm, we use what is called a Monte Carlo algorithm, in which no seed is provided to the code so that each run is a bit different. (The Stata code is given in Table 2.5, and the R code is provided in Table 2.6.) Here we execute the

TABLE 2.5. Stata: Monte Carlo Poisson Code

```
program define poix_sim, rclass   /* poix_sim.ado */
version 10
drop _all
set obs 50000
gen x1 = runiform()
gen x2 = runiform()
gen x3=  runiform()
gen py = rpoisson(exp(1 + 0.75*x1 - 1.25*x2 + .5*x3))
glm py x1 x2 x3, nolog fam(poi)
return scalar sx1 = _b[x1]        /// x1
return scalar sx2 = _b[x2]        /// x2
return scalar sx3 = _b[x3]        /// x3
return scalar sc  = _b[_cons]     /// intercept
return scalar ddisp = e(dispers_s)  /// deviance dispersion
return scalar pdisp = e(dispers_p)  /// Pearson dispersion
end
```

run 500 times, keeping the values of the coefficients and both dispersion statistics. At the conclusion, there are 500 observations of data in memory. Obtaining the mean of the vector of mean values for each model statistic provides us with the Monte Carlo results. The **simulate** command calls and runs **poix_sim**:

```
. simulate mx1 = r(sx1)  mx2 = r(sx2) mx3 = r(sx3) mcon = r(sc)
  mdd = r(ddisp) mpd = r(pdisp) , reps(500) : poix_sim

  ...

. sum
```

Variable	Obs	Mean	Std. Dev.	Min	Max
mx1	500	.7499434	.0086789	.7273185	.7806724
mx2	500	-1.249551	.0089773	-1.275919	-1.223709
mx3	500	.4993433	.0091402	.4705572	.5229337
mcon	500	1.00001	.0082247	.9790608	1.021834
mdd	500	1.104367	.0066602	1.083313	1.124828
mpd	500	1.00027	.006106	.9778052	1.020143

The Monte Carlo results demonstrate that we are modeling a true Poisson model and that the dispersion statistic is 1.0. The deviance dispersion is 1.10, some 10% greater than 1. The results for the value of the Pearson-based dispersion are identical for any "true" Poisson model we test using this method.

Monte Carlo results using R are similar to those for Stata. The dispersion statistic is 0.9999323, nearly identical to 1. Coefficients are nearly correct to the ten-thousandths place. If we had replicated the model 500 times as we did for Stata instead of 100 times, our results would have been even closer:

TABLE 2.6. R: Monte Carlo Poisson Code

```
mysim <- function()
{
 nobs <- 50000
 x1 <- runif(nobs)
 x2 <- runif(nobs)
 x3 <- runif(nobs)
 py <- rpois(nobs, exp(2 + .75*x1 - 1.25*x2 + .5*x3))
 poi <- glm(py ~ x1 + x2 + x3, family=poisson)
   pr <- sum(residuals(poi, type="pearson")^2)
   prdisp <- pr/poi$df.residual
   beta <- poi$coef
   list(beta,prdisp)
}
B <- replicate(100, mysim())
apply(matrix(unlist(B[1,]),3,100),1,mean)
```

```
# Coefficients: intercept and x1, x2, x3
> apply(matrix(unlist(B[1,]),4,100),1,mean)
[1]  1.9998512  0.7499429 -1.2497058  0.5000332

# Dispersion
> mean(unlist(B[2,]))
[1] 0.9999323
```

Analysts use synthetic models for a number of reasons. We learned what a true Poisson model looks like and learned which dispersion statistic to test when checking for possible Poisson model overdispersion. In fact, many analysts still believe that the deviance dispersion is the appropriate test statistic to assess overdispersion. But simulation studies, such as the Monte Carlo runs we employed in this book, clearly demonstrate that the Pearson dispersion statistic is 1.0 for a true Poisson, with greater values indicating overdispersion and lower values underdispersion. The deviance dispersion is biased, tending to display more correlation in the data when there is none. R's **summary** function, when used following **glm**, displays the deviance and residual degrees of freedom.

2.4 Poisson Regression: Modeling Real Data

I will use the well-known German national health registry data set for an example of a Poisson regression. The data were collected for a number of years beginning with 1984, which will be used for the example model. In Stata, the data are saved as **rwm1984**. R users will have to install and load the **msme** or **COUNT** (use capitals) package, then load the **rwm5yr** data, which cover the years 1984–1988. Code will be given to create a data file named **rwm1984**, which is a subset of **rwm5yr**.

The response, or dependent, variable is the number of visits made by a patient to a physician (*docvis*) during the year. *docvis* is the count variable to be explained by other variables selected from the remainder of the data. The selected variables are called explanatory predictors. I'll use *outwork* (1 = patient not working; 0 = patient working) and *age* (25–64) as the two model predictors to explain *docvis*. As in the previous section, I follow model estimation using the **abic** command, which is on the book's web site. The **modelfit** function in the **COUNT** package on CRAN provides the same statistics.

As previously mentioned, knowing one's data is vital to the modeling process. This is particularly important when modeling count data. The response variable is *docvis*. A partial output of a tabulation of the variable can be displayed in Stata as follows:

```
STATA CODE
. use rwm1984
. tab docvis
```

MD visits/year	Freq.	Percent	Cum.
0	1,611	41.58	41.58
1	448	11.56	53.15
2	440	11.36	64.51
3	353	9.11	73.62

```
       4 |         213         5.50        79.12
       5 |         168         4.34        83.45
       6 |         141         3.64        87.09
       7 |          60         1.55        88.64
                   ...          ...          ...
```

Notice the 41.58% zero (0) counts. With a mean of 3.163 and variance of 39.39 (variance = standard deviation squared), the variance far exceeds the mean:

```
. sum docvis

    Variable |        Obs        Mean    Std. Dev.        Min        Max
-------------+--------------------------------------------------------------
      docvis |       3874    3.162881    6.275955          0        121
```

In addition, the expected number of zero (0) counts, based on a Poisson distribution mean of 3.162881, is

```
. di exp(-3.162881)* (3.162881^0)/exp(lnfactorial(0))
.04230369
```

We expect that only 4.23% of the observations in the model have a zero count, but there are over 41% zero counts.

The model we are working with is seriously Poisson overdispersed. In fact, it would be rather surprising if the data came from a Poisson distribution; or alternatively, surprising if the data could be appropriately modeled using Poisson regression.

The two other predictor variables we are using are *outwork*

```
. tab outwork

       1=not |
    working; |
   0=working |       Freq.      Percent        Cum.
-------------+---------------------------------------
           0 |       2,454        63.35        63.35
           1 |       1,420        36.65       100.00
-------------+---------------------------------------
       Total |       3,874       100.00
```

and *age*

```
. sum age

    Variable |       Obs        Mean    Std. Dev.       Min        Max
-------------+---------------------------------------------------------
         age |      3874    43.99587     11.2401        25         64
```

There is no problem with *outwork*, which is fairly balanced, but *age* is a discrete variable consisting of integer values from 25 to 64 with a mean of 44. *age* is a predictor, presumably explaining, in part, if older ages tended to result in a patient visiting the doctor more often during the year. Since *age*

```
. distinct age
             |       Observations
    Variable |     total    distinct
-------------+---------------------
         age |      3874          40
```

has 40 distinct values, and there is no reason to categorize it into separate indicator variables, we enter it into the model as continuous. Most statisticians will *center* a continuous predictor when it starts far from 0, as in this case. We will do the same:

```
. center age, pre(c)    /* to also standardize age, type "stand" as added
                           option */
```

Centering is the process where the mean of the variable is subtracted from every value of the variable. For example, centering provides that

```
age − mean(age)
```

for each age in the model. Using the **center** command, we have named it *cage*. It is important to remember that centering changes only the value of the intercept in the model. All other predictor coefficients and standard errors, and fit statistics, stay the same. We interpret *age* differently when centered, though, as will be seen:

```
. glm docvis outwork cage, fam(poisson) nolog    // mid header deleted

Generalized linear models                 No. of obs      =      3874
Optimization     : ML                     Residual df     =      3871
                                          Scale parameter =         1
```

```
Deviance         =   24190.35807      (1/df) Deviance =   6.249124
Pearson          =   43909.61512      (1/df) Pearson  =   11.34322 <=
                                      AIC             =   8.074027
Log likelihood   =  -15636.39048      BIC             =   -7792.0
---------------------------------------------------------------------
             |                 OIM
    docvis   |    Coef.   Std. Err.     z    P>|z|    [95% Conf. Interval]
-------------+-------------------------------------------------------
   outwork   |  .4079314   .0188447   21.65  0.000    .3709965    .4448663
      cage   |  .0220842   .0008377   26.36  0.000    .0204423    .0237261
     _cons   |  .9380964   .0127571   73.54  0.000    .913093     .9630999
---------------------------------------------------------------------

. abic
AIC Statistic   =    8.074027         AIC*n       = 31278.781
BIC Statistic   =    8.07418          BIC(Stata)  = 31297.566
```

Other predictors in the data could have been used in the preceding model as well, but I selected a model with a single binary and single continuous predictor to keep it simple.

Using the Stata **glm** command, the *nolog* option is used so that the iteration log is not displayed. The intercept value .938 is the linear predictor of all observations in the model when both *outwork* and *age* have values of 0. If *age* were not centered, the intercept would have a value of −.03351697.

It is of primary importance to notice that the Pearson-based dispersion statistic has a value of 11.34, far higher than the ideal value for a Poisson model of 1.0. I determined elsewhere that the model is not apparently overdispersed. Adding an interaction, for example, or squaring the *cage* variable, and so forth, did not result in an elimination of overdispersion. That would have solved the issue upfront. Given the fact that

- there are excessive 0 counts in *docvis* given its mean,
- the variance of *docvis* far exceeds its mean, and
- the dispersion statistic (11.34) is much greater than 1,

it is not unwarranted to conclude that the model is truly overdispersed. The predictors all have *p*-values of under 0.05, and thereby appear to significantly contribute to understanding (and predicting) *docvis*, but *they do not*. The bias resulting from overdispersion – especially such extreme overdispersion – means that the *p*-values tell us nothing about the relationship of the predictor and response. It is imperative that you remember this when

TABLE 2.7. R: Example Poisson Model and Associated Statistics

```
library(COUNT)
data(rwm5yr); rwm1984 <- subset(rwm5yr, year==1984)
cage <- rwm1984$age - mean(rwm1984$age)
summary(poic <-
glm(docvis ~ outwork + cage, family=poisson, data=rwm1984))
pr <- sum(residuals(poic, type="pearson")^2)  # Pearson Chi2
pr/poic$df.residual                           # dispersion statistic
modelfit(poic)
cnt <- table(docvis)
dataf <- data.frame(prop.table(table(docvis) ) )
dataf$cumulative <- cumsum(dataf$Freq)
datafall <- data.frame(cnt, dataf$Freq*100, dataf$cumulative * 100)
datafall
```

constructing and interpreting a model. **An example Poisson model in R is** given in Table 2.7.

Perhaps we can develop a better-fitted model if we add more "significant" predictors. I have heard many analysts claim that this is the case. However, it is not. It may be the case that the "true" model includes certain additional predictors (see Table 2.3) that were excluded from the model and when included enhance the model fit, but it is also possible that added "significant" predictors have little impact on the fit – particularly when the model has been misspecified and should be based on an alternative probability distribution (e.g., negative binomial, PIG, or even a ZIP). It may also be the case that including additional predictors does appear to provide a better fit but that the appearance of better fit is merely superficial. Of course, the added predictors may actually be required for a better fit. However, too many "significant" predictors will tend to overfit the model (i.e., the model may fit the sample data so well that it cannot be used to inform us about other data from the overall population of data). The overfitted model is specific only to these particular data. Also be cautious about not having too many predictors for a given number of observations in the data. A common rule of thumb is that there should be no less than a 10:1 ratio of observations to predictors. Some authors have even argued that 20:1 or greater is preferable. A fairly recent article demonstrated through simulation studies that for logistic and Cox proportional hazards models, which are similar to Poisson models, a more relaxed ratio is often feasible, depending on the structure of the data. A

5:1 ratio or smaller may at times produce a well-fitted model. See Vittinghoff and McCulloch (2006) for details. We explore alternatives in subsequent chapters.

Let us add additional predictors to our model and view the summary statistics:

```
ADDED BINARY PREDICTORS: female, married, and kids
CATEORICAL PREDICTOR:   Educational level (edlevel)  --- a 4-
category predictor.

STATA CODE
. tab edlevel

     Level of |
    education |      Freq.     Percent        Cum.
------------+-----------------------------------
 Not HS grad |      3,152       81.36       81.36
     HS grad |        203        5.24       86.60
   Coll/Univ |        289        7.46       94.06
 Grad School |        230        5.94      100.00
------------+-----------------------------------
       Total |      3,874      100.00
```

I initially used the term *i.edlevel* to tell Stata that *edlevel* is categorical, with the first level as the default reference level, and therefore is excluded from estimation. Displayed in Stata and R,

```
. glm docvis outwork cage female married kids i.edlevel, fam(poi) nolog
or
> summary(tst1 <- glm(docvis ~ outwork + cage + female + married + kids
    + factor(edlevel), family=poisson, data=rwm1984))
```

I found, however, that the second level was not significantly different from the reference level (i.e., its *p*-value was substantially greater than 0.05). I then generated four indicator or dummy variables, excluding both levels 1 and 2 from estimation. This means that a combination of levels 1 and 2 is the reference level (patients who have not attended college or university). A model can then be developed as

```
. tab edlevel, gen(edlevel)  // create indicator vars for each level
                                  of edlevel
. glm docvis outwork cage female married kids edlevel3-edlevel4,
  fam(poi) nolog
```

```
Generalized linear models                No. of obs    =      3874
Optimization      : ML                   Residual df   =      3866
                                         Scale parameter =        1
Deviance       =   23894.1626            (1/df) Deviance =  6.18059
Pearson        =   43855.49193           (1/df) Pearson  = 11.34389
                                         AIC           =  8.000151
Log likelihood  =  -15488.29274          BIC           = -8046.895
------------------------------------------------------------------
             |            OIM
      docvis |   Coef.   Std. Err.     z    P>|z|  [95% Conf. Interval]
-------------+----------------------------------------------------
     outwork |  .2725941  .0215047   12.68  0.000   .2304456   .3147426
        cage |  .0194905  .0009708   20.08  0.000   .0175878   .0213933
      female |  .2596274  .0211399   12.28  0.000   .2181938   .3010609
     married | -.0937657  .022653    -4.14  0.000  -.1381648  -.0493666
        kids | -.1247957  .0222546   -5.61  0.000  -.1684139  -.0811776
     edlevel3 |  -.19083   .0398098   -4.79  0.000  -.2688558  -.1128042
     edlevel4 | -.2594583  .0480192   -5.40  0.000  -.3535742  -.1653424
       _cons |  1.010466  .023941    42.21  0.000   .9635427   1.05739
------------------------------------------------------------------

. abic
AIC Statistic   =    8.000151         AIC*n     = 30992.586
BIC Statistic   =    8.004609         BIC(Stata) = 31042.682
```

Alternatively, I can simply create a new categorical variable with a combined first two levels from *edlevel* and remodel using a new three-level categorical predictor that we define and relabel:

```
. gen elevel = edlevel
. recode elevel 2=1  3=2   4=3
. lab define elevel 1 "HS" 2 "Coll/Univ" 3 "Grad School"
. lab values elevel elevel
. tab elevel

      elevel |    Freq.     Percent      Cum.
-------------+-----------------------------------
         HS |    3,355       86.60      86.60
   Coll/Univ |      289        7.46      94.06
 Grad School |      230        5.94     100.00
-------------+-----------------------------------
      Total |    3,874      100.00

. glm docvis outwork cage female married kids i.elevel, fam(poi) nolog
            <not displayed>
```

TABLE 2.8. R: Change Levels in Categorical Predictor

```
levels(rwm1984$edlevel)                    # levels of edlevel
elevel <- rwm1984$edlevel                  # new variable
levels(elevel)[2] <- "Not HS grad"         # assign level 1 to 2
levels(elevel)[1] <- "HS"                  # rename level 1 to "HS"
levels(elevel)                             # levels of elevel
summary(tst2 <- glm(docvis ~ outwork + cage + female + married + kids
    + factor(elevel), family=poisson, data=rwm1984))
```

R code for creating *elevel* from *edlevel* is shown in Table 2.8:

```
> levels(rwm1984$edlevel)
[1] "Not HS grad" "HS grad"      "Coll/Univ"    "Grad School"

> levels(elevel)
[1] "HS"           "Coll/Univ"    "Grad School"
```

It's clear that even with five "significant" predictors added to the model, the dispersion and AIC and BIC statistics are nearly identical to those of the reduced model with *outwork* and *cage* as predictors. In such a case, *Occam's razor* maintains. Occam's dictum in this context can be addressed as:

> *If two models have the same explanatory power,*
> *the simpler model is preferred.*

Prediction is one of the most important features of a statistical model. We address it in Section 2.6 of this chapter.

2.5 INTERPRETING COEFFICIENTS AND RATE RATIOS

2.5.1 How to Interpret a Poisson Coefficient and Associated Statistics

The Poisson model we constructed in the previous section appears not to fit well. We speculated on what may have given rise to the overdispersion we found in the model. It is clear that we must adjust it in order to discover the best-fitted model for the data. What if the model was well fitted, though? How should the preceding coefficients be interpreted? What about standard errors, and z-values? How do we interpret the statistics that are displayed in Stata, R, and SAS output?

The coefficient β_j in general is the change in the log-count of the response for a one-unit change in the predictor. For a binary predictor such as *outwork*, β is the change in the log-count value of the response when the value of the predictor changes from 0 to 1. For a continuous predictor, the change with respect to the predictor is from the lower of two contiguous values to the next higher; for example, from age 45 to age 46. Note, though, that the change in the response is in terms of log-values (i.e., log-physician visits), not the original values of the response. When a continuous predictor is centered, the reference is its mean value. When a centered variable is part of an interaction, it helps to reduce correlation and is therefore in general desirable. Predictions are not affected. A continuous variable may also be scaled, which entails that the centered variable be standardized.

Let's take another look at the coefficient table for the German health data we have been evaluating. This time, however, I will not center *age* but leave it as it comes in the data – as discrete values from 25 to 64. Stata allows the user to display only the coefficient table by using the *nohead* option:

```
. glm docvis outwork age, fam(poi) nolog nohead
------------------------------------------------------------------------------
             |                 OIM
      docvis |      Coef.   Std. Err.      z    P>|z|     [95% Conf. Interval]
-------------+----------------------------------------------------------------
     outwork |   .4079314   .0188447    21.65   0.000     .3709965    .4448663
         age |   .0220842   .0008377    26.36   0.000     .0204423    .0237261
       _cons |   -.033517   .0391815    -0.86   0.392    -.1103113    .0432773
------------------------------------------------------------------------------
```

It's obvious why most analysts prefer not to interpret Poisson coefficients. It's rather difficult to decipher the import of a log-count. I provided two alternative ways of expressing the generic meaning of the coefficient (see Table 2.9), and other versions can be given as well, but they are all difficult to comprehend or really visualize.

A coefficient provides a way of quantifying the relationship of the predictor to the response based on the model. But its value is only as good as its statistical significance. It has become fairly standard to define .05 as the criterion of statistical significance. In preliminary studies, this criterion is typically relaxed, and in subjects like physics and nanotechnology the relationship is

TABLE 2.9. Interpretation of Poisson Coefficients

generic: The response has a log-count increase of β for a one-unit increase in the value of the predictor. Likewise, the response has a log-count decrease of β for a one-unit decrease in the value of the predictor. Other predictors are held at their mean value.

outwork: Patients who are out of work increase the log-number of visits to a physician by 0.4 compared with a patient who is working, holding *age* at its mean.

age: For each one-year increase in *age*, there is an increase in the expected log-count of visits to the doctor of .022, holding outwork at its mean.

tightened to .01 or even .001. But this is rare. For our purposes, we'll maintain the generally accepted significance level of .05.

I provided the formula for the Poisson Hessian matrix in Chapter 1 and mentioned that the standard errors of the model parameter estimates are obtained as the square root of the diagonal terms of the inverse negative of the Hessian. This can be simplified by having the software being used to model data display the variance–covariance matrix of the model coefficients. That's the negative inverse Hessian.

Following estimation of a model, the variance–covariance matrix can be obtained as

```
STATA CODE
. matrix list e(V)

symmetric e(V)[3,3]
                   docvis:     docvis:     docvis:
                   outwork     age         _cons
docvis:outwork    .00035512
   docvis:age    -4.485e-06    7.018e-07
 docvis:_cons     .00003565   -.00003104   .00153519

. di sqrt(.00035512)      //  standard error of outwork
.01884463

. di sqrt(7.018e-07)      // standard error of age
.00083774

. di sqrt(.00153519)      //  standard error of the intercept
.0391815
```

The values we calculate match the standard errors displayed in the coefficient table. These values are referred to as the *model standard errors*. We define and use other types of standard errors later in this chapter.

A z-value is simply the ratio of the coefficient and associated standard error. For the *outwork* predictor, we have

```
. di 0.4079314 /0.0188447
21.64701
```

The other coefficient/SEs are the same.

The p-value is based on the normal distribution. Since the intercept is the only one of the three z-values having a $p > 0.00$, I'll demonstrate its calculation:

```
. di normal(-0.033517 / 0.0391815)*2
.39231357
```

Confidence intervals are based on the formula

$$\beta_j \pm z^{\alpha/2}se(\beta_j) \tag{2.1}$$

where β_j is the coefficient of an individual predictor. $z^{\alpha/2}$ is a quantile from the normal distribution. For an α of .05, $z = 1.96$. We can therefore calculate the 95% confidence intervals of *outwork* as

```
. di .4079314 -  1.96* .0188447      //  lower confidence interval
.37099579
```

```
. di .4079314 +  1.96* .0188447      //  upper confidence interval
.44486701
```

which perfectly matches the coefficient table confidence intervals.

For models with few observations and models with unbalanced data, coefficients generally fail to approach normality, which is assumed when using standard methods. In fact, many coefficients are not distributed normally, particularly those that are not statistically significant. Many analysts prefer to "model" the standard errors using a likelihood-based method. The default calculation for confidence intervals using R's **glm** function is the use of likelihood-based standard errors that are obtained using the **confint** function. Stata's **pllf** command provides likelihood profiling of standard errors. Only continuous predictors are subject to likelihood profiling:

```
R CODE
> summary(pyq <- glm(docvis ~ outwork + age, family=poisson,
  data=rwm1984))

# Likelihood Profiling of SE
> confint(pyq)

Waiting for profiling to be done...
                2.5 %      97.5 %
(Intercept) -0.1105313 0.04305984
outwork      0.3709909 0.44486231
age          0.0204444 0.02372828

# Traditional Model-based SE
> confint.default(pyq)

                2.5 %      97.5 %
(Intercept) -0.11031129 0.04327726
outwork      0.37099654 0.44486632
age          0.02044229 0.02372611
```

The standard errors differ little in this likelihood profiling, but this is not always the case. Profiling may mean the difference between accepting or rejecting a predictor. In general, *standard errors based on profile likelihood are preferable to traditional model-based standard errors.* We will later find that determining the significance of a predictor by means of a likelihood ratio test is usually preferable to the traditional model-based method. In addition, I will suggest using another type of adjustment to model-based standard errors – the use of robust or sandwich variance estimators (refer to Section 3.4.3).

2.5.2 Rate Ratios and Probability

In order to have a change in predictor value reflect a change in actual visits to a physician, we must exponentiate the coefficient – e^{β_j}. Using Stata's **glm** command, the *eform* option exponentiates the coefficients and confidence intervals of the coefficients. The standard error is calculated using the *delta* *method*, which in this case amounts to $\exp(\beta)*\beta_{SE}$, where the second term is the standard error of the original standard model. First, we exponentiate the coefficients:

```
STATA CODE
#  exponentiated coefficient: rate ratio outwork
. di   exp( _b[outwork])
1.5037041

#  exponentiated coefficient: rate ratio age
. di   exp( _b[age])
1.0223299

* Delta Method: SE outwork rate ratio
. di   exp( _b[outwork]) * _se[outwork]
.02833683

* Delta Method: SE age rate ratio
. di   exp( _b[age]) * _se[age]
.00085643
```

The calculations may be verified by displaying the table of incidence rate ratios:

```
. glm docvis outwork age, fam(poi) nolog nohead eform
---------------------------------------------------------------------------
             |                OIM
      docvis |      IRR   Std. Err.      z    P>|z|    [95% Conf. Interval]
-------------+-------------------------------------------------------------
     outwork |  1.503704   .0283368   21.65   0.000    1.449178    1.560282
         age |   1.02233   .0008564   26.36   0.000    1.020653     1.02401
       _cons |  .9670385     .03789   -0.86   0.392    .8955553    1.044227
---------------------------------------------------------------------------
```

The header statistics of the exponentiated model are identical to the standard table of coefficients. Using R, we can obtain exponentiated coefficients by running *exp(coef(poi1))*.

The incidence rate ratio (IRR) indicates the ratio of the rate of counts between two ascending contiguous levels of the response. For this example, we interpret the IRRs for each predictor as in Table 2.10.

TABLE 2.10. Interpretation of Rate Ratios

Patients who were out of work in 1984 had one-and-a-half times more visits to
a physician during 1984 than patients who were working. *Age* is held
constant.

Patients visited a physician some 2.2% more often with each year older in age.

If *age* is centered, the result will be interpreted as such.

For each year difference from the mean of *age*, there is an associated 2.2% decrease
or increase in the mean number of physician visits made by German patients in
1984. *Outwork* is held constant.

Other ways of expressing the same thing as in the table may also be given.
In particular, rate ratios can be expressed as probabilities or likelihoods in
the sense that we can also interpret *outwork* as

> *Out-of-work patients were some one-and-a-half times more likely (or
> more probable) to visit a physician in 1984 than were working patients.*

If we think of our data as a random sample from a greater population of
data that can be described (or was generated) by a distribution with specific
unknown parameters, it is possible to interpret *outwork* as:

> *Out-of-work patients are about one-and-a-half times more likely to
> visit a doctor than are working patients;*

or

> *Those with a job see a doctor half as much as those who are out of
> work.*

For the predictor *age*, we can express the rate ratio as:

> *For each year older, Germans are likely to see a doctor some two-and-
> a-half percent more often.*

Most analysts prefer to exponentiate coefficients and interpret parameter
estimates as rate ratios. Always remember, though, that such an interpretation
always includes a comparison of two levels of a predictor. For a binary

TABLE 2.11. R: Poisson Model – Rate Ratio Parameterization

```
library(COUNT)
data(rwm5yr); rwm1984 <- subset(rwm5yr, year==1984)
summary(poi1 <- glm(docvis ~ outwork + age, family=poisson,
data=rwm1984))
pr <- sum(residuals(poi1, type="pearson")^2)  # Pearson Chi2
pr/poi1$df.residual                     # dispersion statistic
poi1$aic / (poi1$df.null+1)             # AIC/n
exp(coef(poi1))                         # IRR
exp(coef(poi1))*sqrt(diag(vcov(poi1)))  # delta method
exp(confint.default(poi1))              # CI of IRR
```

variable, the ratio is between $x = 1$ and $x = 0$. For categorical variables, the ratio is between the level of interest and the reference level. For continuous variables, it is the ratio of the higher value of the variable to the value of the variable at the next lowest level (e.g., age 45 to 44).

Note that the preceding Poisson model may also be estimated using Stata's **poisson** command. It uses full MLE to obtain model results, which are the same as what is produced using **glm**. R has MLE versions of Poisson regression using the **COUNT, msme, and gamlss** packages, among others. To display the available functions in the **COUNT** package (or other packages), use the command

```
ls("package:COUNT")
```

Using R, we may obtain the same results by running the code in Table 2.11 in the R script editor.

2.6 EXPOSURE: MODELING OVER TIME, AREA, AND SPACE

The basic Poisson distribution assumes that each count in the distribution occurs over a small interval of time. However, we often count events over a period of time or over various areas. In fact, this is what we mean by a rate – a count of events within a given time period, area, or volume. With respect to time, we may want to count how many events occur each day over a period of a month, or over any period of time; for example, how many distinct sunspots were counted each year over a 20-year or longer period or how many sales of a particular type of product occur each month over the course of a year. The rate of counts, μ, is calculated as the number of events counted divided by the period of time that counting occurs, and likewise for counts per area. We may count 10 events occurring in area A and only 7 in area B, but if area A is half the size of area B, the rate of counts of events is greater in area B. The same is the case for volumes of space: how many supernovae are counted in an area of 100 cubic light-years in our Milky Way galaxy. We may have to adjust over both time and area, or time and volume, but most studies do not entail such a double adjustment.

> *Statisticians use an **offset** with a model to adjust for counts of events over time periods, areas, and volumes. The model is sometimes referred to as a proportional intensity model.*

We briefly addressed the rate parameterization of the Poisson model in Chapter 1. Although μ is sometimes said to be an intensity or rate parameter, it is such only when thought of in conjunction with a constant coefficient, t. The rate parameterization of the Poisson PDF can be expressed as

$$f(y;\mu) = \frac{e^{-t\mu}(t\mu)^y}{y!} \tag{2.2}$$

where t represents the length of time, or exposure, during which events or counts uniformly occur. t can also be thought of as an area or volume in which events uniformly occur, each associated with a specific count. For instance, when using a Poisson model with disease data, $t^*\mu$ can be considered the rate of disease incidence in specified geographic areas, each of which may differ from other areas in population. Likewise, the incidence rate of hospitalized bacterial pneumonia patients can be compared across counties within the state. A count of such hospitalizations divided by the population size of the county, or by the number of total hospitalizations for all diseases, results in the *incidence rate ratio* (IRR) for that county. When $t = 1$, the model is understood to apply to individual counts without a consideration of size. Technically, basic Poisson distribution assumes that each count in the distribution occurs over a small interval of time – so small that it consists of a single count. Size or period is not a consideration. If the periods, areas, or volumes in which events occur are the same for the entire study, an offset is not needed. On the other hand, where unequal periods of time, area, or volume (TAV) occur in the model, an offset must be given. The key notion is that events can be considered as entering or being in an area or period of time independently of other events. They are uniformly distributed in each t.

TABLE 2.12. R: Poisson with Exposure

```
data(fasttrakg)
summary(fast <- glm(die ~ anterior + hcabg + factor(killip),
                    family=poisson,
                    offset=log(cases),
                    data=fasttrakg))
exp(coef(fast))
exp(coef(fast))*sqrt(diag(vcov(fast)))
exp(confint.default(fast))
modelfit(fast)
```

Many commercial software applications indicate exponentiated Poisson coefficients as incidence rate ratios by using the acronym IRR. IRR is also used with exponentiated negative binomial and PIG coefficients.

When employing a rate parameter over TAV to a Poisson model, analysts enter the natural log of t as an offset into the estimating algorithm. The fitted value is expressed as

$$\mu = \exp(x\beta + \log(t)) \qquad (2.3)$$

which is the same as

$$\exp(x\beta) = \frac{\mu}{t}, \qquad \mu = t\exp(x\beta) \qquad (2.4)$$

$\log(t)$ is entered into the estimating algorithm as a constant.

An example of a rate-parameterized Poisson model is provided. The data are from the Canadian National Cardiovascular Disease registry called FASTRAK. They have been grouped by covariate patterns from individual observations. The response is *die*, which is a count of the number of deaths of patients having a specific pattern of predictors. Predictors are *anterior*, which indicates if the patient has had a previous anterior myocardial infarction; *hcabg*, if the patient has a history of having had a CABG (coronary artery bypass grafting) procedure; and *killip* class, a summary indicator of the cardiovascular health of the patient, with increasing values indicating increased disability. The number of observations sharing the same pattern of covariates is recorded in the variable case. This value is log-transformed and entered into the model as an offset. In Stata, the offset is specified as an option by using either the *lnoffset* or *exposure* options, both of which automatically log the offset

TABLE 2.13. R: Interpretation of Poisson Exposure Model

Patients having an anterior site heart attack are twice as likely to die than if the damage was to another area of the heart.

Patients with a history of having a CABG procedure are twice as likely to die than if they did not have such a procedure.

Patients having a killip 2 status are two-and-a-half times more likely to die than if they have level 1 killip level status (no perceived problem). Those at level 3 are 3 times more likely to die, and those at level 4, which is experiencing a massive heart attack, are 12 times more likely to die than those with no apparent heart problems.

variable before entering it into the estimation algorithm. Stata developers now prefer *exposure* instead of *lnoffset*.

The data are in **fasttrakg** and are given in the following display. Since the data are grouped, they are easier to observe as a whole. The R **fasttrakg** data file is in the **COUNT** package (see Table 2.12 for R code for Poisson exposure; see Table 2.13 for R interpretation of a Poisson exposure model).

```
STATA CODE
. use fasttrakg,clear
. l, nolab    // list of numeric values for each observation and predictor
     +-------------------------------------------------------------------+
     | die   cases   anterior   hcabg   killip   kk1   kk2   kk3   kk4 |
     |-------------------------------------------------------------------|
  1. |   5      19          0       0        4     0     0     0     1 |
  2. |  10      83          0       0        3     0     0     1     0 |
  3. |  15     412          0       0        2     0     1     0     0 |
  4. |  28    1864          0       0        1     1     0     0     0 |
  5. |   1       1          0       1        4     0     0     0     1 |
     |-------------------------------------------------------------------|
  6. |   0       3          0       1        3     0     0     1     0 |
  7. |   1      18          0       1        2     0     1     0     0 |
  8. |   2      70          0       1        1     1     0     0     0 |
  9. |  10      28          1       0        4     0     0     0     1 |
 10. |   9     139          1       0        3     0     0     1     0 |
     |-------------------------------------------------------------------|
 11. |  39     443          1       0        2     0     1     0     0 |
 12. |  50    1374          1       0        1     1     0     0     0 |
 13. |   1       6          1       1        3     0     0     1     0 |
 14. |   3      16          1       1        2     0     1     0     0 |
 15. |   2      27          1       1        1     1     0     0     0 |
     +-------------------------------------------------------------------+
```

```
. glm die anterior hcabg kk2-kk4, nolog fam(poi) eform exposure(cases)

Generalized linear models              No. of obs      =         15
Optimization     : ML                  Residual df     =          9
                                       Scale parameter =          1
Deviance         =   10.93195914       (1/df) Deviance =   1.214662
Pearson          =   12.60791065       (1/df) Pearson  =   1.400879
                                       AIC             =    4.93278
Log likelihood   = -30.99584752        BIC             =  -13.44049

-----------------------------------------------------------------------
             |                OIM
         die |      IRR   Std. Err.      z    P>|z|   [95% Conf. Interval]
-------------+---------------------------------------------------------
    anterior |  1.963766   .3133595    4.23   0.000    1.436359    2.684828
       hcabg |  1.937465   .6329708    2.02   0.043    1.021282    3.675546
         kk2 |  2.464633   .4247842    5.23   0.000     1.75811    3.455083
         kk3 |  3.044349   .7651196    4.43   0.000     1.86023    4.982213
         kk4 |  12.33746   3.384215    9.16   0.000    7.206717    21.12096
       _cons |  .0170813   .0024923  -27.89   0.000    .0128329    .0227362
   ln(cases) |         1   (exposure)
-----------------------------------------------------------------------

. abic
AIC Statistic   =      4.93278       AIC*n       =  73.991692
BIC Statistic   =     5.566187       BIC(Stata)  =  78.239998
```

The dispersion at first may appear as relatively low at 1.40, but given a total observation base of 5388, the added 40% overdispersion may represent a lack of model fit. We delay this discussion until later in this chapter.

2.7 Prediction

We have previously described how predicted Poisson probabilities may be generated. However, one of the foremost uses of prediction rests with the ability of predicting expected counts based on predictor values. The values do not have to be the same as those that actually were used to estimate the model parameters. For example, using the German health data example, we may calculate the expected number of doctor visits for a working 50-year-old patient based on the table of coefficient values displayed as

```
glm docvis outwork age, nolog fam(poi) nohead
-----------------------------------------------------------------------
             |              OIM
      docvis |    Coef.   Std. Err.      z    P>|z|    [95% Conf. Interval]
-------------+---------------------------------------------------------
     outwork |  .4079314   .0188447   21.65   0.000    .3709965   .4448663
         age |  .0220842   .0008377   26.36   0.000    .0204423   .0237261
       _cons |  -.033517   .0391815   -0.86   0.392   -.1103113   .0432773
-----------------------------------------------------------------------
```

Given that a working patient has an *outwork* value of 0 and coefficient values are signified as _b[ ], we may obtain the linear predictor for a 50-year-old working patient as

```
. di _b[outwork]*0 + _b[age]*50 - .033517
1.0706929
```

The expected count is calculated by exponentiating the Poisson linear predictor:

```
. di exp(1.0706929)
2.9174003
```

Therefore, on the basis of the model, we expect or predict that a 50-year-old working German patient visited a doctor three times during 1984. If we had more predictors, or a better-fitting model, we would be more certain that our prediction was correct. As it is, we have previously demonstrated that the model is rather severely overdispersed.

Predicted counts and their 95% confidence interval may be obtained from:

```
STATA
#delimit ;
poisson docvis outwork age ; predict mu ; predict eta, xb ;
predict se_eta, stdp ; gen low = eta - invnormal(0.975) * se_eta ;
gen up  = eta + invnormal(0.975) * se_eta ; gen lci = exp(low);
gen uci = exp(up) ; sort mu;
twoway (line lci mu uci eta, lpattern( dash 1 dash 1)),
     ytitle("Predicted Count and   95% CI"); #delimit cr
```

```
R
myglm <- glm(docvis ~ outwork + age, family=poisson, data=rwm1984)
lpred <- predict(myglm, newdata=rwm1984, type="link", se.fit=TRUE
up <- lpred + 1.96*lpred$se.fit); lo <- lpred - 1.96*lprd$se.fit)
eta <- lpred$fit ; mu <- myglm$family$linkinv(eta)
upci <- myglm$family$link(up); loci <- myglm$family$link(lo)
```

In the next chapter, we extend the analysis of predictions to making comparisons between observed and predicted, or expected, counts.

2.8 POISSON MARGINAL EFFECTS

Marginal effects are used now by most econometricians. In fact, some in this area believe that marginal effects should be reported in place of coefficients for models such as Poisson and negative binomial regression. They are also used for the other models we discuss in this book. What are they?

- *Marginal effects* pertain only to continuous predictors.
- *Discrete change* or partial effects are used for binary and categorical predictors.

I believe that these statistical methods are important to research in other disciplines as well, so it is wise to be aware of what they do.

A marginal effect relates a continuous predictor to the predicted probability of the response variable. Other predictors in the model are held at their mean,

TABLE 2.14. R: Marginal Effects at Mean

```
==================================================
library(COUNT)
data(rwm5yr); rwm1984 <- subset(rwm5yr, year==1984)
summary(pmem <- glm(docvis ~ outwork + age, family=poisson,
        data=rwm1984))
mout <- mean(rwm1984$outwork); mage <- mean(rwm1984$age)
xb <- coef(pmem)[1] + coef(pmem)[2]*mout + coef(pmem)[3]*mage
dfdxb <- exp(xb) * coef(pmem)[3]
mean(dfdxb)

==================================================
> mean(dfdxb)
[1] 0.06552846
```

or sometimes median, values. The basic interpretation of marginal effects relates to

> *how the probability of the count response changes with a one-unit change in the value of the continuous predictor.*

We discuss two types of marginal effects:

- the marginal effect taken at the mean value of the predictor and
- the average marginal effect taken at the mean of the predicted counts.

2.8.1 Marginal Effect at the Mean

For count models in general, and in particular Poisson and negative binomial models, the marginal effect at the mean is defined as

$$\text{MEmean} = \exp(x_i' \beta_k) \beta_k$$

Using our **rwm1984** data, we calculate the marginal effect of *age*, holding *outwork* at its mean:

```
STATA CODE
. margins, dydx(age) atmeans

Conditional marginal effects                      Number of obs   =    3874
Model VCE     : OIM

Expression    : Predicted mean docvis, predict()
dy/dx w.r.t.  : age
at            : outwork        =     .3665462 (mean)
                age            =     43.99587 (mean)
------------------------------------------------------------------------------
             |            Delta-method
             |     dy/dx   Std. Err.      z    P>|z|     [95% Conf. Interval]
-------------+----------------------------------------------------------------
         age |  .0655285   .0024232    27.04   0.000     .060779     .0702779
------------------------------------------------------------------------------
```

The marginal effect of *age*, .0655, may be understood to mean:

> *The number of visits made to a doctor increases by 6.6% for each year of age, when* outwork *is set at its mean value.*

Or:

> *At the average of age, an extra year of age is associated with a 0.066
> increase in doctor visits per year when outwork is set at its mean value.*

There is no standard R function yet that calculates marginal effects at the
mean, or average marginal effects. We must therefore write R script to determine the results (see Table 2.14).

```
> mean(dfdxb)
[1] 0.06552846
```

This is the same value obtained using the Stata **margins** command.

2.8.2 Average Marginal Effects

Average marginal effects are defined as

$$\beta_k \bar{y}$$

Using Stata, we may calculate them as

```
STATA CODE
. margins, dydx(age)

Average marginal effects                        Number of obs  =    3874
Model VCE     : OIM

Expression    : Predicted mean docvis, predict()
dy/dx w.r.t. : age
-----------------------------------------------------------------------
             |            Delta-method
             |     dy/dx    Std. Err.      z    P>|z|     [95% Conf. Interval]
----+------------------------------------------------------------------
age |    .0698497    .0027237    25.64   0.000     .0645113    .0751881
-----------------------------------------------------------------------
```

> *The above-average marginal effects tell us that there are on average
> 0.07 more doctor visits given each additional year of patient age, with
> outwork at its mean value.*

Or:

> *For each additional year of age, there are on average 0.07 additional
> doctor visits, with outwork at its mean value.*

Using R, we need to execute the following code. Note that the calculated statistic is identical to what we calculated using the **margins** command:

```
R CODE
> mean(rwm1984$docvis) * coef(pmem)[3]
      age
0.06984968
```

2.8.3 Discrete Change or Partial Effects

Discrete change is used to evaluate the change in predicted probability of the response when a binary predictor changes values from 0 to 1.

To determine the partial effects for *outwork*, the predictor in Stata must specifically be made a factor variable. Following the modeling of the data – not displayed because the *quietly* option is used – the **margins** command can be used to obtain the correct effect. To reiterate, for a discrete change, the binary or categorical predictor must be specifically factored in the regression:

```
STATA CODE
. qui glm docvis i.outwork age,  fam(poi)
. margins, dydx(outwork) atmean

Conditional marginal effects                    Number of obs  =   3874
Model VCE    : OIM

Expression   : Predicted mean docvis, predict()
dy/dx w.r.t. : 1.outwork
at           : 0.outwork      =    .6334538 (mean)
               1.outwork      =    .3665462 (mean)
               age            =    43.99587 (mean)
------------------------------------------------------------------------
             |            Delta-method
             |   dy/dx    Std. Err.     z    P>|z|    [95% Conf. Interval]
--------+---------------------------------------------------------------
1.outwork |  1.287021  .0625526   20.58   0.000    1.16442    1.409622
------------------------------------------------------------------------
Note: dy/dx for factor levels is the discrete change from the base level.

R CODE
> summary(pmem <- glm(docvis ~ outwork + age, family=poisson,
          data=rwm1984))
> mu0 <- exp(pmem$coef[1] + pmem$coef[3]*mage)
```

```
> mu1 <- exp(pmem$coef[1] + pmem$coef[2] + pmem$coef[3]*mage)
> pe <- mu1 - mu0
> mean(pe)
[1] 1.287021
```

Average partial effects or discrete change uses the same method as for average marginal effects, except for the factoring of *outwork*:

```
STATA CODE
. margins, dydx(outwork)

Average marginal effects                         Number of obs   =   3874
Model VCE     : OIM
Expression    : Predicted mean docvis, predict()
dy/dx w.r.t.  : 1.outwork
----------------------------------------------------------------------
             |            Delta-method
             |    dy/dx   Std. Err.     z    P>|z|   [95% Conf. Interval]
-----------+----------------------------------------------------------
 1.outwork |  1.327171  .0635453   20.89   0.000   1.202624   1.451718
----------------------------------------------------------------------
Note: dy/dx for factor levels is the discrete change from the base level.
```

```
R CODE
> summary(pmem <- glm(docvis ~ outwork + age, family=poisson,
          data=rwm1984))
> bout = coef(pmem)[2]
> mu = fitted.values(pmem)
> xb = pmem$linear.predictors
> pe_out = 0
> pe_out = ifelse(rwm1984$outwork == 0, exp(xb + bout)-exp(xb), NA)
> pe_out = ifelse(rwm1984$outwork == 1, exp(xb)-exp(xb-bout),pe_out)
> mean(pe_out)
[1] 1.327171
```

There are several other ways to relate predictors and the response; for example, as elasticities and semielasticities. If you intend to use marginal effects with your modeling project, I suggest you check the possibility of parameterizing the relationship of continuous predictors to the response as elasticities or even semielasticities. See Hilbe (2011) for a more extensive analysis of marginal and partial effects, and alternative statistics.

2.9 Summary

We have covered quite a bit of statistical material in this chapter, starting from the distributional assumptions on which the Poisson model is based and then identifying and testing for apparent overdispersion as distinct from real overdispersion. We then discussed constructing and interpreting Poisson models. A true synthetic Poisson model was created, identifying aspects of the model output that characterize well-fitted models. This discussion was followed by modeling a poorly fitted real-data Poisson model, also evaluating its characteristics. Finally, we explained the meaning of two varieties of marginal and partial effects and described how they are created as well as how they are interpreted.

A Poisson model is usually a good place to start when modeling count data, which may have a number of explanatory predictors. The following chapter expands on the meaning of overdispersion and testing for it. We also address standard fit tests as they relate to count models. By the conclusion of the chapter, you should have a good understanding of the basics of modeling count data. What will remain are models that have been designed to adjust for specific types of extradispersion; for instance, negative binomial regression, Poisson inverse Gaussian regression, and so forth.

## CHAPTER 3

Testing Overdispersion

SOME POINTS OF DISCUSSION

- How is overdisperion recognized?
- What are some of the foremost tests to determine whether a Poisson model is overdispersed?
- What is scaling? What does it do to a count model?
- Why should robust standard errors be used as a default?
- What is a quasi-likelihood model?

This chapter can be considered a continuation of Chapter 2. Few real-life Poisson data sets are truly equidispersed. Overdispersion to some degree is inherent in the vast majority of Poisson data. Thus, the real question deals with the amount of overdispersion in a particular model – is it statistically sufficient to require a model other than Poisson? This is one of the foremost questions we address in this chapter, together with how we assess fit and then adjust for the lack of it.

3.1 BASICS OF COUNT MODEL FIT STATISTICS

Most statisticians consider overdispersion the key problem when considering count model fit. That is, when thinking of the fit of a count model, an analyst typically attempts to evaluate whether a count model is extradispersed – which usually means overdispersed. If there is evidence of overdispersion in a Poisson model, the problem then is to determine what gives rise to it. If we can determine the cause, we can employ the appropriate model to use on the data.

Analysts have used a variety of tests to determine whether the model they used on their data actually fits. The earliest fit test used with Poisson regression, as a member of the GLM family, is called the deviance goodness-of-fit test. The test is based on the deviance statistic, which for many years was the standard statistic used for IRLS algorithm convergence and was a standard statistic displayed in GLM model output. Now most IRLS algorithms use either the deviance or log-likelihood as the basis of convergence, but most still display the deviance statistic in model results.

The deviance is based on the log-likelihood function, which earlier we saw is a reparameterization of a probability function. Note, however, that it is possible to construct a log-likelihood function based on the variance of the distribution being modeled, independent of any probability function. In fact, a likelihood may be developed regardless of any underlying PDF. This is called quasi-likelihood estimation, which we address later in our discussion. For now, we assume that the log-likelihood is a true one, based on a PDF. In the case of Poisson regression, the log-likelihood function is based on the Poisson PDF.

The deviance is defined as the difference between a saturated log-likelihood and full model log-likelihood. The saturated log-likelihood is calculated by changing every μ (*mu*) in the function to a y. This represents a situation in which there is a parameter for every observation in the model. It indicates a model with a perfect, but uninformative, fit:

$$D = 2 \sum_{i=1}^{n} \{\mathcal{L}(y_i; y_i) - \mathcal{L}(\mu_i; y_i)\} \tag{3.1}$$

The Poisson log-likelihood function is given as

$$\mathcal{L}(\mu; y) = \sum_{i=1}^{n} \{y_i \log(\mu) - \mu_i - \log(y_i!)\} \tag{3.2}$$

The saturated log-likelihood function is $y \log(y) - y - \log(y!)$. Subtracting this function by equation (3.2) results in

$$D = 2 \sum_{i=1}^{n} y_i \log\left(\frac{y_i}{\mu_i}\right) - (y_i - \mu_i) \tag{3.3}$$

Note that log(y!), the normalization term that provides for the function to sum to 1, cancels. In fact, the normalization term for every GLM model PDF cancels when calculating the respective deviance statistic. Since the normalization terms can get rather complicated for some distributions, we can see why the deviance was preferred as the basis of algorithm convergence – especially in the 1970s and 1980s, when computing speed was slow.

The deviance goodness-of-fit (GOF) test is based on the view that the deviance is distributed as Chi2. The Chi2 distribution has two parameters – the mean and scale. For the deviance GOF, this is the deviance statistic and residual degrees of freedom, which is the number of observations less predictors in the model, including the intercept and interactions. If the resulting Chi2 p-value is less than 0.05, the model is considered well fit. Employing the deviance GOF test on the modeled German health data yields the following model:[1]

```
. glm docvis outwork age, fam(poi) nolog // non-numeric part of header deleted
Generalized linear models                   No. of obs      =       3874
Optimization     : ML                       Residual df     =       3871
                                            Scale parameter =          1
Deviance         =   24190.35807            (1/df) Deviance =   6.249124
Pearson          =   43909.61512            (1/df) Pearson  =   11.34322
                                            AIC             =   8.074027
Log likelihood   =  -15636.39048            BIC             =   -7792.01
-----------------------------------------------------------------------
             |                 OIM
      docvis |     Coef.   Std. Err.      z    P>|z|   [95% Conf. Interval]
-------------+---------------------------------------------------------
     outwork |   .4079314   .0188447    21.65   0.000   .3709965   .4448663
         age |   .0220842   .0008377    26.36   0.000   .0204423   .0237261
       _cons |   -.033517   .0391815    -0.86   0.392  -.1103113   .0432773
-----------------------------------------------------------------------
.(scalar dev=e(deviance)
. scalar df=e(df)
. di " deviance GOF "" D="dev " df="df " p-value= " chiprob(df, dev)
 deviance GOF  D=24190.358 df=3871 p-value= 0
```

---

[1] For pedagogical purposes, I am only using a single binary and continuous predictor. There are more predictors in the data that may significantly contribute to the fit of the model. In fact, the added predictors do not eradicate the overdispersion inherent in the data. Later, we will further discuss the strategy of adding predictors to effect an optimally fitted model.

TABLE 3.1. R: Deviance Goodness-of-Fit Test

```
> library(COUNT); data(rwm5yr); rwm1984 <- subset(rwm5yr, year==1984)
> mymod <-glm(docvis ~ outwork + age, family=poisson, data=rwm1984)
> mymod
          .   .   .

Coefficients:
(Intercept)        outwork            age
   -0.03352        0.40793        0.02208


Degrees of Freedom: 3873 Total (i.e. Null);   3871 Residual
Null Deviance:      25790
Residual Deviance: 24190          AIC: 31280


> dev<-deviance(Model); df<-df.residual(Model)
> p_value<-1-pchisq(dev,df)
> print(matrix(c("Deviance GOF"," ","D",round(dev,4),"df",df, "p_value",
p_value), ncol=2))
     [,1]               [,2]
[1,] "Deviance GOF" " "
[2,] "D"                "24190.3581"
[3,] "df"               "3871"
[4,] "p-value"          "0"
```

The deviance is 24190.35807 and residual degrees of freedom is 3871. These are both saved postestimation statistics that can be used for just this sort of purpose. R code and results are given in Table 3.1. Note that it is common to see the symbol G or G^2 used for the deviance test. I have used D instead.

Again, for the deviance GOF test, as well as the Pearson GOF tests that follow, a Chi2 $p < 0.05$ indicates that the model is considered well fit. More accurately, statistically speaking,

> with a $p < 0.05$, the deviance GOF test indicates that we can reject the hypothesis that the model is not well fitted.

We also assume that the model has not been misspecified and that the model may appear to fit well by chance alone, as is the case for all such tests. Also, since it was first proposed, statisticians have discovered that many models appearing to be well fitted on the basis of the deviance test in fact poorly fit the data. Recall that a *p*-value of 0.05 accepts that we will be mistaken 1 out of 20 times on average.

If the value of D is very large, then we can generally be safe in rejecting the goodness of the model fit:

> *Deviance is in effect a measure of the distance between the most full or complete (saturated) model we can fit and the proposed model we are testing for fit.*

The smaller the distance, or deviance, between them, the better the fit. The test statistic evaluates whether the value of the deviance, for a specific size of model, is close enough to that of the saturated model that it cannot be rejected as not fitting.

Many older texts on generalized linear models use a deviance test to compare nested models. A nested model is one that has one or more predictors dropped from the main model. The *p*-value for the difference in deviance between the two models tells us whether the predictors we excluded are statistically important to the modeling of the data. A *p*-value under 0.05 advises us that they are important. Parameters may be dropped as well, comparing similar models that have and do not have that parameter. But the test is usually given to models with and without predictors. In either case, the smaller model is said to be nested within the larger model. We can also view the test as evaluating if we should add a predictor to the model. If the difference in the deviance statistic is minimal when adding a predictor or predictors, we may conclude that they will contribute little of statistical value to the model and therefore should not be incorporated into the final model. The logic of this type of reasoning is well founded, but there is no well-established *p*-value to quantify the significance of the differences in deviances. There is, though, for comparing log-likelihoods, and it is called a likelihood ratio test.

Many analysts use the Pearson Chi2 statistic, Pearson Chi2/rdof, in place of the deviance GOF statistic. We will not consider using the Pearson Chi2 statistic for a GOF test, though. It appears to produce biased results. However, its true value comes from its use in defining overdispersion, in particular Poisson overdispersion. The reason for this is that the Pearson Chi2 statistic is nothing other than the squared residuals weighted or adjusted by the model variance, and summed across all observations in the model:

$$\chi^2 = \sum_{i=1}^{n} \frac{(y_i - \mu_i)^2}{V(\mu_i)} \tag{3.4}$$

The Pearson statistic may also be viewed as the sum of the squared Pearson residuals as defined later in this chapter. The dispersion statistic of the Poisson model is defined as the Pearson Chi2 statistic divided by the residual degrees of freedom. To reiterate from the previous discussion, the *residual degrees of freedom* is the number of observations in the model less the number of predictors, including the intercept and interactions.

The sum of squared residuals is an absolute raw measure (squaring eliminates negative values) of the difference in observed versus predicted model counts, adjusted by both the variance and size of the model. Adjustment is made by dividing the squared residuals by the product of the variance and residual degrees of freedom. The result is the dispersion statistic, which should have a value of 1 if there is no unaccounted variability in the model other than what we expect based on the variance function. Values greater than 1 indicate an overdispersed model; values less than 1 are underdispersed. We discuss the dispersion statistic in much greater detail throughout this book.

Some statisticians have used the deviance dispersion as the basis for scaling standard errors. However, as we will find in this book, simulation studies indicate that

> *when scaling standard errors, the Pearson dispersion statistic better captures the excess variability in the data, adjusting model standard errors in such a manner as to reflect what the standard errors would be if the excess variability were not present in the data.*

In R, Poisson models with scaled standard errors are called *quasipoisson*:

> *A Pearson dispersion in excess of 1.0 indicates likely Poisson model overdispersion. Whether the overdispersion is significant depends on (1) the value of the dispersion statistic, (2) the number of observations in the model, and (3) the structure of the data; for example, if the data are highly unbalanced.*

A command exists in Stata to calculate the deviance and Pearson Chi2 statistics and associated *p*-values for the associated GOF tests. However, it must be used following the **poisson** command, which is a full maximum likelihood estimation algorithm that can be used for Poisson models in place

TABLE 3.2. R: Function to Calculate Pearson Chi2 and Dispersion Statistics

```
P__disp <- function(x) {
    pr <- sum(residuals(x, type="pearson")^2)
    dispersion <- pr/x$df.residual
    cat("\n Pearson Chi2 = ", pr ,
        "\n Dispersion   = ", dispersion, "\n")
}
```

of **glm**. I'll run the **poisson** command quietly (no displayed results), then use the "**estat gof**" command to produce both GOF tests:

```
STATA CODE
. qui poisson docvis outwork age
. estat gof

        Deviance goodness-of-fit  =   24190.36
        Prob > chi2(3871)         =      0.0000

        Pearson goodness-of-fit   =   43909.61
        Prob > chi2(3871)         =      0.0000
```

In R, I can do the same by using

```
R CODE
> mymod <-glm(docvis ~ outwork + age, family=poisson, data=rwm1984)
> pr <- sum(residuals(mymod, type="pearson")^2)    # get Pearson Chi2
> pchisq(pr, mymod$df.residual, lower=F)            # calc p-value
[1] 0
> pchisq(mymod$deviance, mymod$df.residual, lower= F)  # calc p-vl
[1] 0
```

CALCULATING THE DISPERSION STATISTIC

I created the **P__disp.r** function to use following R's **glm** and **glm.nb** functions. **P__disp.r** calculates the Pearson Chi2 and associated dispersion statistics (see Table 3.2). It can be pasted into the script editor and run before using it or can be loaded into memory from a site on your computer where it is saved. It is also part of the **COUNT** package on CRAN. If **COUNT** is installed and loaded into memory, you can use the function anytime. A deviance test can be created instead by replacing the second line to obtain the deviance.

For example, to load the function from where it is saved in the c:\\Rfiles directory or folder, type the following:

```
> source("c:\\Rfiles\\P__disp.r")  # use folder where you keep R files
> P__disp(mymod)
```

On the other hand, if **COUNT** is loaded, simply use the function. As an example, we will calculate the Pearson Chi2 and dispersion statistics for the Poisson model of the German health data found in Table 3.1. Recall that R's **glm** function does not display either the Pearson Chi2 or dispersion statistics. They must be calculated. For the negative binomial model, we use the **nbinomial** function in place of **glm.nb**. **nbinomial** by default displays the Pearson Chi2 statistic and dispersion, plus the summary of Pearson residuals, in addition to statistics provided by **glm.nb**. Note also that the **glm** function with a *quasipoisson* "family" displays the Pearson dispersion statistic but not the Pearson Chi2:

```
R  CODE
> library(COUNT)
> data(rwm5yr)
> rwm1984 <- subset(rwm5yr, year==1984)
> mymod <-glm(docvis ~ outwork + age, family=poisson, data=rwm1984)
> P__disp(mymod)

 Pearson Chi2 =  43909.62
 Dispersion   =  11.34322
```

3.2 OVERDISPERSION: WHAT, WHY, AND HOW

Not all overdispersion is real; apparent overdispersion can sometimes be identified and the model amended to eliminate it. In Section 2.2, I indicated what constitutes apparent overdispersion and how to eradicate it. Now we want to determine its causes and how it is to be tested.

In Table 3.3, we list the major concerns about overdispersion in general (from Hilbe 2007a).

3.3 TESTING OVERDISPERSION

The first set of tests to give a model when there is evidence of overdispersion are those listed in Table 3.3. These are possible remedies for apparent

---

TABLE 3.3. Overdispersion: What, Why, and How

1. What is overdispersion? Overdispersion in a Poisson model occurs when the variance of the response is greater than its mean. Overdispersion in general is the occasion when the observed variance in a model is greater than its expected variance.
2. What causes overdispersion? Overdispersion is caused by a positive correlation between responses or by an excess variation between response probabilities or counts. Overdispersion also arises when there are violations in the distributional assumptions of the data as well as when there is proneness in the data (i.e., when earlier events cause or influence the existence of subsequent events).
3. Why is overdispersion a problem? Overdispersion may cause standard errors of the estimates to be underestimated (i.e., a variable may appear to be a significant predictor when it is in fact not significant).
4. How is overdispersion recognized? A model may be overdispersed if the value of the Pearson Chi2 (χ^2) divided by the degrees of freedom (dof) is greater than 1.0. The quotient is called the dispersion. Small amounts of overdispersion are of little concern; however, if the dispersion statistic is greater than 1.25 for moderate-sized models, then a correction may be warranted. Models with large numbers of observations may be overdispersed with a dispersion statistic of 1.05.
5. What is apparent overdispersion? How may it be corrected? Apparent overdispersion occurs when
   (a) the model omits important explanatory predictors,
   (b) the data include outliers,
   (c) the model fails to include a needed interaction term or terms,
   (d) a predictor needs to be transformed to another scale (e.g., log or square root), or
   (e) the assumed linear relationship between the response and the link function and predictors is mistaken (i.e., the link is misspecified). In the case of count models, this criterion is not applicable for most circumstances.

---

overdispersion, not specific tests of overdispersion. If equidispersion can result in a Poisson model by simply adding an interaction term, or a predictor that was previously neglected, then we do not need to test the model for future overdispersion. The important thing to remember, though, is that for the Poisson model, almost all violations of distributional assumptions, or other situations that result in a poorly fitted model, also result in overdispersion, and in some cases underdispersion:

- Add appropriate predictor
- Construct required interactions
- Transform predictor(s)
- Transform response
- Adjust for outliers
- Use correct link function

For example, let's see what happens if there is a missing predictor in a Poisson model. To be sure that a predictor is really needed, we must use a synthetic model for testing. In Section 2.3, we created a synthetic Poisson model with three predictors. Refer to the displayed output, noting the values of the parameter estimates and dispersion statistic, which is .9989237. If we exclude $x2$ from the model, we now have a model appearing as

```
. glm py x1 x3, nolog fam(poi)

Generalized linear models             No. of obs      =        50000
Optimization     : ML                 Residual df     =        49997
                                      Scale parameter =            1
Deviance       =   73496.70743        (1/df) Deviance =     1.470022
Pearson        =   68655.76474        (1/df) Pearson  =     1.373198
                                      AIC             =     4.116233
Log likelihood =  -102902.8168        BIC             =    -467459.7
-----------------------------------------------------------------------
             |               OIM
        py |     Coef.   Std. Err.      z    P>|z|     [95% Conf. Interval]
-------------+---------------------------------------------------------------
        x1 |  .7513276   .0090462    83.05   0.000     .7335973    .7690579
        x3 |  .4963757   .0090022    55.14   0.000     .4787317    .5140196
     _cons |   .444481   .0075083    59.20   0.000      .429765    .4591969
-----------------------------------------------------------------------
```

The predictor coefficients are close to their true values; that is, close to the Poisson distribution parameter values we specified when creating the data. The intercept is about half its true value. However, we would not normally know that the value of the intercept differs from the parameters used in creating the data. The fact that the predictor *p*-values all approximate 0.000 appears to indicate that the model is well fitted. Many researchers are mislead in this manner.

The value of the Pearson-based dispersion statistic is key to providing the analyst with a test of fit. The Poisson model assumes that the mean and variance of the response term are approximately equal. This equidispersion is indicated when the dispersion statistic has a value of 1. Excluding $x2$ from the model causes the dispersion statistic to be inflated to a value of 1.37, indicating overdispersion.

In this case, we know that adding $x2$ back into the model will result in a well-fitted model, and the dispersion statistic will have a value approximating 1. But with real empirical data for which we are attempting to develop a well-fitted model, we will not know in advance the true parameter values. We do know, however, that if a Poisson model dispersion statistic does not approximate 1.0, then it is not equidispersed, and not well fitted. We can use the tests discussed in this and other texts to help identify the source of the extradispersion, and if successful attempt to fix the model. But it may be the case that the data are simply not Poisson – that we must amend the basic Poisson model or even use another model on the data.

To reiterate, just because the predictors of a Poisson model are all significant does not in itself indicate that the model is well fitted. We must check the dispersion statistic. If the dispersion does not approximate a value of 1.0, the model is not a well-fitted Poisson model, regardless of the significance of the predictors. We cannot rely on the model to appropriately inform us about the response. Predictions based on an ill-fitted model and conclusions based on the coefficients and their exponentiations are biased. An alternative model needs to be constructed.

If an analyst is confronted with a Poisson model having a dispersion statistic greater than 1.0, I suggest that they first check the six remedies we gave earlier to determine whether applying one results in an equidispersed Poisson model. The model may be only apparently over- or underdispersed. If such adjustment fails to result in a dispersion of approximately 1.0, try to determine what caused the extradispersion and identify a model suited for it.

There are tests for Poisson overdispersion other than checking the dispersion. These will be examined next.

3.3.1 Score Test

The concept of overdispersion is central to the understanding of both Poisson and negative binomial models. Nearly every application of the negative

TABLE 3.4. R: z-Score Test

```
library(COUNT); data(rwm5yr); rwm1984 <- subset(rwm5yr, year==1984)
summary(poi <- glm(docvis ~ outwork + age, family=poisson, data=rwm1984))
mu <-predict(poi, type="response")
z <- ((rwm1984$docvis - mu)^2 - rwm1984$docvis)/ (mu * sqrt(2))
summary(zscore <- lm(z ~ 1))
```

binomial is in response to perceived overdispersion in a Poisson model. I have thus far only touched on the problem of ascertaining whether indicators of overdispersion represent real overdispersion in the data or only apparent. Apparent overdispersion can usually be accommodated by various means in order to eradicate it from the model. However, real overdispersion is a problem affecting the reliability of both the model parameter estimates and fit in general.

In the previous section, I showed one manner in which overdispersion could be detected in a Poisson model. We'll address other methods in the next chapter. However, two related yet well-used tests are at times provided in commercial software applications. These are the score and Lagrange multiplier tests. (R code for each is given in Tables 3.4 and 3.5, respectively.)

A score test to evaluate whether the amount of overdispersion in a Poisson model is sufficient to violate the basic assumptions of the model may be defined as

$$z = \frac{(y - \mu)^2 - y}{\mu\sqrt{2}} \tag{3.5}$$

The test is *post hoc* (i.e., it is executed after the data have been modeled). Using the **rwm1984** data as earlier defined, we first model the data using Poisson

TABLE 3.5. R: Lagrange Multiplier Test

```
obs <- nrow(rwm1984)    # continue from Table 3.2)
mmu <- mean(mu); nybar <- obs*mmu; musq <- mu*mu)
mu2 <- mean(musq)*obs)
chival <- (mu2 - nybar)^2/(2*mu2); chival)
pchisq(chival,1,lower.tail = FALSE))
```

regression. Then we predict μ, calculate z from the preceding formula, and regress z using a basic linear regression:

```
. glm docvis outwork age,  fam(poi) nolog nohead
----------------------------------------------------------------------
            |                 OIM
     docvis |    Coef.    Std. Err.      z     P>|z|     [95% Conf. Interval]
------------+---------------------------------------------------------
    outwork |  .4079314   .0188447    21.65   0.000     .3709965    .4448663
        age |  .0220842   .0008377    26.36   0.000     .0204423    .0237261
      _cons |  -.033517   .0391815    -0.86   0.392    -.1103113    .0432773
----------------------------------------------------------------------
. predict mu
. gen double z=((docvis-mu)^2-docvis)/ (mu*sqrt(2))
. regress z
     Source |       SS       df       MS              Number of obs =    3874
------------+------------------------------           F(  0,  3873) =    0.00
      Model |        0        0        .              Prob > F      =       .
   Residual | 16012865     3873   4134.4862           R-squared     = 0.0000
------------+------------------------------           Adj R-squared = 0.0000
      Total | 16012865     3873   4134.4862           Root MSE      =    64.3
----------------------------------------------------------------------
          z |    Coef.    Std. Err.      t     P>|t|     [95% Conf. Interval]
------------+---------------------------------------------------------
      _cons |  7.30804    1.033073    7.07    0.000     5.282622    9.333459
----------------------------------------------------------------------
```

The z score test is 7.3, with a t-probability of < 0.0005. z tests the hypothesis that the Poisson model is overdispersed; it evaluates whether the data are Poisson or negative binomial. This example indicates that the hypothesis of no overdispersion is rejected (i.e., that it is likely that real overdispersion exists in the data).

The test is based on two assumptions:

- The data set on which the test is used is large.
- z is t-distributed.

The preceding z-test result indicates that there is overdispersion in the data. Other z tests have been constructed, and you will find a variety of alternatives in the literature. The most popular alternative is perhaps an *auxiliary regression test*, for which the equation is the same as equation (3.5) but without the adjustment of $\sqrt{2}$, or simply 1.4142 (which is easy to remember using a mnemonic, such as 14–14, or 1 with a decimal of 41–42). The analyst can test whether the overdispersion in a Poisson model can be adjusted for by using an NB1 or NB2 model. I do not discuss it further since there are more powerful methods you should use. It is important, though, to be aware of the existence of the test should you see it in a study report.

3.3.2 Lagrange Multiplier Test

The Lagrange multiplier test is a Chi2 test. It can be defined as

$$\chi^2 = \frac{\left(\sum_{i=1}^n \mu_i^2 - n\bar{y}_i\right)^2}{2\sum_{i=1}^n \mu_i^2} \tag{3.6}$$

with one degree of freedom. Again, using Stata commands to calculate the statistic, we have

```
STATA CODE
. summ docvis, meanonly     /* solving for Lagrange Multiplier */
. scalar nybar = r(sum)
. gen double musq = mu*mu
. summ musq, meanonly
. scalar mu2 = r(sum)
. scalar chival = (mu2-nybar)^2/(2*mu2)

. display "LM value = " chival _n "P-value = " chiprob(1,chival)
LM value = 11582.694
P-value = 0
```

With one degree of freedom, the test appears to be significant – the hypothesis of no overdispersion is again rejected:

> *A Chi2 statistic of 3.84 has a p-value of 0.05.*
> *Higher values of Chi2 result in lower p-values.*
> *For example, a Chi2 of 7.5 has a p of 0.00617.*

The preceding inset should be memorized. Chi2 tests permeate statistics.[2]

---

[2] Recall that 95% confidence intervals ($p = .05$) are defined as $\beta \pm 1.96*\beta(se)$. This defines a one-sided value of the normal PDF. 1.96 is in fact a $p = .025$. For

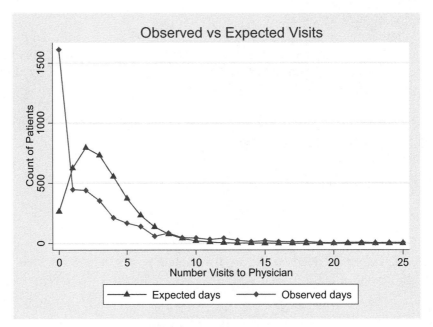

FIGURE 3.1. Observed versus expected doctor visits.

3.3.3 Chi2 Test: Predicted versus Observed Counts

A majority of analysts consider the most important test of fit for a count model to be an analysis of the difference between observed and expected counts across the full range of counts in the data (see Figure 3.1). Tables 3.6 and 3.7 provide the Stata and R code, respectively, to create a table of observed versus predicted (expected) counts following model estimation. The code for each may be amended so that you may incorporate it into your own research projects.

the default meaning of a *p*-value (i.e., two-sided), we want this value so that we encompass .025 on both the left and right sides of the normal distribution: .025*2 = .05.

```
STATA    . di normal(1.96)        . di -invnormal(.025)
             .9750021                 1.959964
R        > pnorm(1.96)           > -qnorm(.025)
         [1] 0.9750021              [1] 1.959964
```

TABLE 3.6. Stata do File – Observed versus Predicted Counts

```
use rwm1984, clear
qui glm docvis outwork age, fam(poisson)
predict mu
count
gen nobs = e(N)
local i 0
local newvar "pr'i'"
* Predicted probability at each day
 while 'i' <=25 {
   local newvar "pr'i'"
   qui gen 'newvar' =  poissonp(mu, 'i')
   local i = 'i' + 1
  }
quietly gen cnt = .
quietly gen observ = .
quietly gen expect = .
local i 0
*: Observed and expected docvis
while 'i' <=25 {
   local obs = 'i' + 1
   replace cnt = 'i' in 'obs'
   tempvar obser
   gen 'obser' = 'e(depvar)' =='i'  /* (docvis=='i') */
   sum 'obser'
   replace observ = r(mean)* nobs in 'obs'
   sum pr'i'
   replace expect = r(mean)* nobs in 'obs'
   local i = 'i' + 1
}
*: Preparation for table
gen byte count = cnt
gen diff = observ - expect
drop cnt pr0-pr25 nobs mu
list count observ expect diff in 1/21
lab var expect "Expected days"
lab var observ "Observed days"
label var count "Number visits to Physician"
twoway scatter expect observ count, c(l l) ms(T d) ///
     title(Observed vs Expected visits) ytitle(Count of patients)
```

TABLE 3.7. R: Poisson Model with Ancillary Statistics

```
library(COUNT); data(rwm5yr); rwm1984 <- subset(rwm5yr, year==1984)
summary(poi1 <- glm(docvis ~ outwork + age, family=poisson, data=rwm1984))
pr <- sum(residuals(poi1, type="pearson")^2)   # Pearson Chi2
pr/poi1$df.residual                      # dispersion statistic
poi1$aic / (poi1$df.null+1)              # AIC/n
exp(coef(poi1))                          # IRR
exp(coef(poi1))*sqrt(diag(vcov(poi1)))   # delta method
exp(confint.default(poi1))               # CI of IRR
modelfit(poi1)                           # same as Stata abic
sd(rwm1984$docvis)^2                     # observed variance
xbp <- predict(poi1)                     # xb, linear predictor
mup <- exp(xbp)                          # mu, fitted Poisson
mean(mup)                                # expected variance: mean=variance
# Table of observed vs expected counts
rbind(obs=table(rwm1984$docvis)[1:18],
    exp = round(sapply(0:17, function(x)sum(dpois(x, fitted(poi1))))))
meany <- mean(rwm1984$docvis)            # mean docvis
expect0 <- exp(-meany)*meany^0/exp(log(factorial(0))) #expected prob of 0
zerodays <- (poi1$df.null+1) *expect0   # expected zero days
obs=table(rwm1984$docvis)[1:18] #observed number values in each count0-17
# expected each count
exp = round(sapply(0:17, function(x)sum(dpois(x, fitted(poi1)))))
chisq.test(obs, exp)                     # ChiSq test if obs & exp from same pop
```

Note: Expected value based on mean of *docvis*. Different from Stata.

```
     +----------------------------------------+
     | count   observe      expect       diff |
     |----------------------------------------|
  1. |    0      1611     264.7923   1346.208  |
  2. |    1       448      627.102   -179.102  |
  3. |    2       440     796.0545  -356.0545  |
  4. |    3       353     731.5981  -378.5981  |
  5. |    4       213     554.6024  -341.6024  |
     |----------------------------------------|
  6. |    5       168     373.3741  -205.3741  |
  7. |    6       141     232.9783  -91.97826  |
  8. |    7        60     137.5237  -77.52374  |
  9. |    8        85     77.19441   7.805588  |
 10. |    9        47     41.09616   5.903843  |
              .       .        .
 20. |   19         6     .0028615   5.997139  |
     |----------------------------------------|
 21. |   20         4     .0008157   3.999184  |
```

```
                    < partial output displayed >
R CODE
> sd(rwm1984$docvis)^2                    # observed variance
[1] 39.38

> mean(mup)                               # expected variance:
mean=variance
[1] 3.16

> rbind(obs=table(rwm1984$docvis)[1:18], exp = round(sapply(0:17,
function(x) sum(dpois(x, fitted(poi1))))))

        0    1    2    3    4    5    6    7   8   9  10 11 12 13 14 15 16 17
obs 1611 448  440  353  213  168  141   60  85  47  45 33 43 25 13 21 13 11
exp  265 627  796  732  555  373  233  138  77  41  21 10  4  2  1  0  0  0

          Pearson's Chi-squared test
data:  obs and exp
X-squared = 258, df = 240, p-value = 0.2027
```

For the Poisson model, the observed variance is 39.38, whereas the expected variance is 3.16, indicating substantial overdispersion. In addition, note the vastly greater number of observed zero counts compared with what we expect given the mean of *docvis*. We will see why we need to explicitly adjust the model as a result of this fact. Without adjustment by the fitting of explanatory predictors, we should have expected 164 of the 3874 patients not to visit a physician during 1984, based on the Poisson distribution. A fitted model predicts that 265 patients failed to visit a physician. In fact, 1611 (41.58%) patients never went to a physician during the year. There are far more zero visits in the data than allowed using the Poisson model, which without adjustment assumes that only 164 (4.2%) patients, given a mean of 3.16 days, did not need to visit a physician. Another model needs to be used to better fit the number of physician visits during 1984. Again, an excess of zero counts is a common reason for having an overdispersed Poisson model. The Chi2 test is sometimes used as a fit test to determine whether the predicted number of counts come from the same population as the observed values for each count, from 0 to a specified upper value. Here the null hypothesis cannot be rejected and the two cannot be significantly separated.

Using Stata, we can determine the observed and expected values for 0 counts as follows:

```
STATA CODE
. count                   // number of counts in model
 3874

. count if docvis==0      // number of 0 counts
 1611

. di 1611/3874            // percentage of 0 counts in data
.41584925

. di exp(-3.162881) * 3.162881^0 / exp(lnfactorial(0))  // predicted
prob of 0's.
.04230369
```

To calculate how many 0 counts are expected, we multiply the number of observations in the model (3874) by 0.0423, giving 164, which is far fewer than the 1611 actually observed in the data.

3.4 METHODS OF HANDLING OVERDISPERSION

3.4.1 Scaling Standard Errors: Quasi-count Models

Scaling of standard errors was the first method used to deal with overdispersion in binomial and count response models. The method replaces the W, or model weight, in the IRLS algorithm when β's (beta's) are calculated

$$\beta = (X'WX)^{-1}X'Wz \qquad (3.7)$$

with the inverse square root of the dispersion statistic. Scaling by the Pearson dispersion statistic entails estimating the model, abstracting the dispersion statistic, and multiplying the model standard errors by the square root of the dispersion, and then running one additional iteration of the algorithm, but as

$$\beta = (X'W_d X)^{-1}X'W_d z \qquad (3.8)$$

Scaling in effect adjusts the model standard errors to the value that would have been calculated if the dispersion statistic had originally been 1.0. The Pearson-based dispersion statistic should always be used to assess count model overdispersion; binomial models may use either the deviance or Pearson dispersion, but I suggest the Pearson.

An example will demonstrate how an overdispersed model can have the standard errors adjusted, providing the user with a more accurate indication of the true standard errors. I will use the **medpar** data that are in the **COUNT** and **msme** packages on CRAN and on this book's web site. The **medpar** data consist of 1991 Arizona Medicare in-patient (hospital) data collected for a particular disease. The key variables for our model are follows:

Response:	*los*	length of stay
Predictors:	*hmo*	1 = member of a Health Maintenance Organization (HMO); 0 = private pay
	white	1 = identifies as white; 0 = other
	type1	1 = elective admission (reference level)
	type2	1 = urgent admission
	type3	1 = emergency admission

The data are modeled as

```
. glm los hmo white type2 type3, fam(poi) nolog
```

Generalized linear models			No. of obs	=	1495
Optimization	: ML		Residual df	=	1490
			Scale parameter =		1
Deviance	=	8142.666001	(1/df) Deviance =		5.464877
Pearson	=	9327.983215	(1/df) Pearson =		6.260391
			AIC	=	9.276131
Log likelihood	=	-6928.907786	BIC	=	-2749.057

los	Coef.	OIM Std. Err.	z	P>|z|	[95% Conf.	Interval]
hmo	-.0715493	.023944	-2.99	0.003	-.1184786	-.02462
white	-.153871	.0274128	-5.61	0.000	-.2075991	-.100143
type2	.2216518	.0210519	10.53	0.000	.1803908	.2629127
type3	.7094767	.026136	27.15	0.000	.6582512	.7607022
_cons	2.332933	.0272082	85.74	0.000	2.279606	2.38626

The Pearson *Chi2* dispersion is an extremely high 6.26, especially considering the relatively large number of observations. For example, based on the original standard error for *hmo* of .023944, we may calculate a scaled standard error as sqrt(6.260391) * .023944 = .05990974. Scaled standard errors are calculated for this model as

```
. glm los hmo white type2 type3, fam(poi) nolog scale(x2)

Generalized linear models                    No. of obs      =        1495
Optimization     : ML                        Residual df     =        1490
                                             Scale parameter =           1
Deviance      =   8142.666001                (1/df) Deviance =    5.464877
Pearson       =   9327.983215                (1/df) Pearson  =    6.260391
                                             AIC             =    9.276131
Log likelihood =  -6928.907786               BIC             =   -2749.057

------------------------------------------------------------------------------
             |                 OIM
        los  |      Coef.   Std. Err.      z    P>|z|     [95% Conf. Interval]
-------------+----------------------------------------------------------------
        hmo  |  -.0715493    .0599097    -1.19   0.232    -.1889701    .0458715
      white  |   -.153871    .0685889    -2.24   0.025    -.2883028   -.0194393
      type2  |   .2216518    .0526735     4.21   0.000     .1184137    .3248899
      type3  |   .7094767    .0653942    10.85   0.000     .5813064     .837647
      _cons  |   2.332933    .0680769    34.27   0.000     2.199505    2.466361
------------------------------------------------------------------------------

(Standard errors scaled using square root of Pearson X2-based dispersion.)

R:   quasipoisson
=========================================================
summary(poiql <- glm(los ~ hmo + white + hmo + factor(type),
                     family=quasipoisson, data=medpar))
=========================================================
```

The R *quasipoisson* family option is aimed to adjust for overdispersion in Poisson models, but it is simply scaling the standard errors using the method described here. In order to obtain the same confidence intervals as Stata, SAS, and other packages, use the *confint.default(poiql)* option. Profile confidence intervals are also displayed for the *quasipoisson option*.

Remember that this method is post hoc, meaning that a standard model is estimated, followed by a calculation of the Pearson dispersion and scaling term, and another iteration of the regression, but with the scaled standard errors in place of model-based standard errors and CIs. By-hand versus *quasipoisson* R code for calculating scaled standard errors and confidence intervals is provided in Table 3.8.

A table of incidence rate ratio statistics can be produced and displayed by exponentiating the model coefficients and appropriately adjusting associated

TABLE 3.8. R: Scaling SE medpar **Data**

```
library(COUNT); data(medpar); attach(medpar)
summary(poi <- glm(los ~ hmo + white + factor(type),
    family=poisson, data=medpar))
confint(poi)
# profile confidence interval
pr <- sum(residuals(poi,type="pearson")^2)          # Pearson statistic
dispersion <- pr/poi$df.residual; dispersion              # dispersion
sse <- sqrt(diag(vcov(poi))) * sqrt(dispersion); sse          # model SE
# OR
poiQL <- glm(los ~ hmo + white + factor(type), family=quasipoisson,
    data=medpar)
coef(poiQL); confint(poiQL)                    # coeff & scaled SEs
modelfit(poiQL)                          # AIC,BIC statistics
```

statistics. The *eform* option generates such output using Stata **glm** command, and the *irr* option with the **poisson** and **nbreg** commands. The R function *exp(coef(model_name))* is used following **glm** and **glm.nb**. A Stata table of rate ratios, with scaled standard errors, can be displayed using the code

```
. glm los hmo white type2 type3, nolog fam(poi) eform scale(x2) nohead
```

```
----------------------------------------------------------------------------
         |              OIM
   los   |     IRR    Std. Err.      z     P>|z|    [95% Conf.  Interval]
---------+------------------------------------------------------------------
   hmo   |  .9309504   .0557729   <= -1.19   0.232   .8278113    1.04694
 white   |  .8573826   .0588069      -2.24   0.025   .7495346    .9807484
 type2   |  1.248137   .0657437       4.21   0.000   1.12571     1.383878
 type3   |  2.032927   .1329416      10.85   0.000   1.788373    2.310923
----------------------------------------------------------------------------
(Standard errors scaled using square root of Pearson X2-based dispersion)
```

Recall that I mentioned earlier that the standard error of the incidence rate ratio (IRR) is not directly based on a model variance–covariance matrix. Rather, standard errors for IRRs are calculated using the *delta* method. Applied to this situation, the scaled standard errors of the IRR are calculated by multiplying the IRR by the scaled model SE. For *hmo*, we have .9309504 * .0599097 = .05577296, which we observe is correct.

It needs to be emphasized that the parameter estimates remain unaffected when standard errors are scaled. When a Poisson (or negative binomial)

TABLE 3.9. medpar **Data: Poisson Model Interpretation**

HMO members have 7% fewer days in the hospital than private-pay patients, holding the other predictor values at their mean.

Patients identifying themselves as white have 15% fewer days in the hospital than nonwhite patients, holding other predictor values at their mean.

Urgent admissions stay in the hospital 25% longer than elective admissions, holding other predictor values at their mean.

Emergency admissions stay in the hospital twice as long as elective admissions, holding other predictor values at their mean.

model is moderately overdispersed, scaling can be an effective way to adjust standard errors to values that would be the case if the model was not overdispersed. Scaling is an easy, quick-and-dirty method of adjusting standard errors for overdispersion. However, when data are highly correlated or clustered, model slopes or coefficients usually need to be adjusted as well. Scaling does not accommodate that need, but simulations demonstrate that it is quite useful for models with little to moderate overdispersion.

3.4.2 Quasi-likelihood Models

Quasi-likelihood (QL) methods were first developed by Wedderburn (1974). They are based on GLM principles but allow parameter estimates to be calculated based only on a specification of the mean and variance of the model observations without regard to those specifications originating from a member of the single-parameter exponential family of distributions. Further generalizations to the quasi-likelihood methodology were advanced by Carroll and Rupert (1981) and Nelder and Pregibon (1987). Called extended quasi-likelihood (EQL) methods, they were designed to evaluate the appropriateness of the QL variance in a model. However, EQL models take us beyond the scope of our discussion. Quasi-likelihood models, though, are important for understanding extensions to the Poisson and negative binomial models we consider later.

Quasi-likelihood methods allow us to model data without explicit specification of an underlying log-likelihood function. Rather, we begin with mean and variance functions, which are not restricted to the collection of functions defined by single-parameter exponential family members, and abstract backward to the implied log-likelihood function. Since this implied log-likelihood is not derived from a probability function, we call it quasi-likelihood or quasi-log-likelihood instead. The quasi-likelihood, or the derived quasi-deviance function, is then used, for example, in an IRLS algorithm to estimate parameters just as for GLMs when the mean and variance functions are those from a specific member of the exponential family.

The quasi-likelihood is defined as

$$Q(y; \mu) = \int_y^\mu \frac{y - \mu}{\phi V(\mu)} \tag{3.9}$$

In an enlightening analysis of leaf-blotch data, the quasi-deviance was applied by Wedderburn using the logit link and a "squared binomial" variance function $\mu^2 (1 - \mu)^2$. However, the same logit could also have been specified with traditional exponential family variance functions. In the case of the Poisson, we see that by taking the integral of $(y - \mu)/\mu$ from μ to y with respect to μ, the resultant equation is the Poisson log-likelihood but without the final $\ln(y!)$ normalizing term. The normalizing term is what ensures that the sum of the probabilities over the probability space adds to unity. The negative binomial (NB2) log-likelihood function can be similarly abstracted using the variance function $\mu + \alpha\mu^2$.

The manner in which quasi-likelihood methodology is typically brought to bear on overdispersed Poisson data is to multiply the variance μ by some constant scale value, indicated as ψ.

The fact that the variance function is multiplied by a constant changes the likelihood, or the deviance function, by dividing it by the scale. It is the next stage in amending the Poisson variance function to adjust for overdispersion. The same **medpar** data will be used as an example. In this case, we enter the Pearson dispersion statistic that we obtained from the base or standard Poisson model as the variance multiplier:

```
QUASI-LIKELIHOOD: VARIANCE MULTIPLIER

. glm los hmo white type2 type3, fam(poi) nolog disp(6.260391) irls
Generalized linear models              No. of obs      =        1495
Optimization     : MQL Fisher scoring  Residual df     =        1490
                   (IRLS EIM)          Scale parameter =    6.260391
Deviance      =   1300.664128          (1/df) Deviance =    .8729289
Pearson       =    1490.00008          (1/df) Pearson  =           1
Quasi-likelihood model with dispersion: 6.260391   BIC    = -9591.059
```

```
--------------------------------------------------------------------------
            |                 EIM
   los |      Coef.   Std. Err.        z     P>|z|     [95% Conf. Interval]
-------------+------------------------------------------------------------
    hmo |  -.0715493   .0095696     -7.48   0.000    -.0903054   -.0527932
  white |   -.153871    .010956    -14.04   0.000    -.1753444   -.1323977
  type2 |   .2216518   .0084138     26.34   0.000     .2051611    .2381424
  type3 |   .7094767   .0104457     67.92   0.000     .6890035    .7299499
  _cons |   2.332933   .0108742    214.54   0.000      2.31162    2.354246
--------------------------------------------------------------------------
```

Extra variation is dampened from the variance when multiplying it by the value of the dispersion, 6.260391, which was given in the output of the earlier scaled model. This is the same as dividing the model standard error by the square root of the dispersion. For example, the model standard error of *hmo* is .023944. Dividing, we have

```
. di .023944/sqrt(6.260391)
.00956965
```

which is the standard error of *hmo* for the quasi-likelihood Poisson model, as we just observed. Note that the Pearson-dispersion value of this QL model is now 1.0. Also, compare to scaling, for which the model standard error is multiplied by the square root of the dispersion.

Compare the summary statistics of this model with the standard Poisson model applied to the same data and the model scaled by the dispersion. The deviance statistic is substantially less than that of the standard and scaled models, which will always have the same values except for the standard errors and related statistics (p, z, CI). The deviance of the standard Poisson is 8242.666, and for the quasi-likelihood model it's 1300.664, indicating a much better fit. By fit in this case we mean more accurate predictor standard errors and p-values. The quasi-likelihood model is not a true likelihood model, and hence the standard errors are not based on a correct model-based Hessian matrix. (R code for quasi-likelihood Poisson standard errors is given in Table 3.10.)

TABLE 3.10. R: Quasi-likelihood Poisson Standard Errors

```
poiQL <- glm(los ~ hmo+white+type2+type3, family=poisson, data=medpar)
summary(poiQL)
pr <-sum(residuals(poiQL, type="pearson")^2 )
disp <- pr/poiQL$df.residual                    # Pearson dispersion
se <-sqrt(diag(vcov(poiQL)))
QLse <- se/sqrt(disp); QLse
```

TABLE 3.11. Implementation of Robust or Empirical Standard Errors

- Estimate the model.
- Calculate the linear predictor, $x\beta$.
- *Calculate the score vector:* $g' = g(\beta; x) = x\partial LL(x\beta)/\partial x\beta) = ux$.
- Calculate the dof adjustment: $n/(n-1)$.
- *Combine terms:* $V(\beta) = V(n/(n-1)\Sigma u^2 x' x)V$.
- Replace the model variance–covariance matrix with a robust estimator: an additional iteration with a new matrix.

Note that with the Stata command, the **irls** option is required to use this method of adjusting standard errors, which, again, are much tighter than scaled standard errors.

This method is rarely used for making adjustments, but the method of multiplying the Poisson variance by some value runs through much of this book. It is important to keep it in mind.

3.4.3 Sandwich or Robust Variance Estimators

Unlike the standard variance estimator, $-H(\beta)^{-1}$, a robust variance estimator adjusts standard errors for correlation in the data. That is, robust standard errors should be used when the data are not independent, perhaps gathered over different households, hospitals, schools, cities, litters, and so forth. Robust variance estimators have also been referred to as sandwich variance estimators or heteroskedastic robust estimators. Other names for this method of adjusting standard errors are Huber or White standard errors, or empirical standard errors. Huber (1967) was the first to discuss this method, which was later independently discussed by White (1980) in the field of econometrics. Robust variance estimators are implemented in a postestimation procedure according to the schema outlined in Table 3.11. Again, their foremost function is to adjust standard errors for data that are not independent. Many statisticians argue that robust standard errors (Table 3.12 **gives R code for the medpar** model) should be the default standard errors for all count response regression models. Others prefer to employ profile likelihood standard errrors as the default. In fact, most researchers use regular model standard errors as the default and employ robust and profile techniques when it is clear that model assumptions have been violated.

We will find that robust variance estimators are quite robust, hence the name, when modeling overdispersion in count response models. They also

TABLE 3.12. R: Robust Standard Errors of medpar **Model**

```
library(sandwich)
poi <- glm(los ~ hmo + white + factor(type), family=poisson, data=medpar)
vcovHC(poi)
sqrt(diag(vcovHC(poi, type="HC0")))    # final HC0 = H-C-zero
# Clustering
poi <- glm(los ~ hmo + white + factor(type), family=poisson, data=medpar)
library(haplo.ccs)
sandcov(poi, medpar$provnum)
sqrt(diag(sandcov(poi, medpar$provnum)))
```

play an important role when interpreting the Poisson or negative binomial parameter estimates as risk ratios.

An example using the same **medpar** data is displayed, but I use the i-dot prefix for creating a factor predictor instead of having indicator or dummy variables for the nonreference levels. It makes no difference in the calculations; only the display differs. The same holds true for R. An analyst can create separate indicator variables for each nonreference level, or use the **factor**() function, with the first level being the default reference. In SAS, the top or highest level is the default reference. Of course, there are other schema used when factoring, as learned in introductory statistics courses, but I will not be concerned with them in this book. In fact, the dummy or indicator approach to factoring categorical variables is by far the most common method used for modeling counts:

```
POISSON WITH ROBUST VARIANCE ESTIMATOR

. glm los hmo white i.type, fam(poi) vce(robust) nolog

Generalized linear models                No. of obs      =        1495
Optimization     : ML                    Residual df     =        1490
                                         Scale parameter =           1
Deviance      =  8142.666001             (1/df) Deviance =    5.464877
Pearson       =  9327.983215             (1/df) Pearson  =    6.260391
                                         AIC             =    9.276131
Log pseudolikelihood = -6928.907786      BIC             =   -2749.057
------------------------------------------------------------------------
            |             Robust
     los |     Coef.   Std. Err.      z    P>|z|     [95% Conf. Interval]
------+-----------------------------------------------------------------
      hmo |  -.0715493   .0517323   -1.38   0.167    -.1729427    .0298441
    white |   -.153871   .0833013   -1.85   0.065    -.3171386    .0093965
```

```
        |
   type |
      2 |    .2216518    .0528824     4.19   0.000     .1180042    .3252993
      3 |    .7094767    .1158289     6.13   0.000     .4824562    .9364972
  _cons |    2.332933    .0787856    29.61   0.000     2.178516     2.48735
--------------------------------------------------------------------------
```

When robust variance estimators are applied to this type of quasi-likelihood model, we find that the effect of the robust variance overrides the adjustment made to the standard errors by the multiplier. It is as if the initial quasi-likelihood model had not been estimated in the first place.

Robust variance estimators can also be applied to models consisting of clustered or longitudinal data. Many data situations take this form. For instance, when gathering treatment data on patients throughout a county, it must be assumed that treatments given by individual providers are more highly correlated within each provider than between providers. Likewise, in longitudinal data, treatment results may be recorded for each patient over a period of time. Again, it should be assumed that results are more highly correlated within each patient record than between patients. Data such as these are usually referred to as panel data. Robust variance adjustments of some variety must be applied to the data to reflect the fact that observations are not independent.

Modified sandwich variance estimators or robust cluster variance estimators provide standard errors that allow inference that is robust to within-group correlation but assume that clusters of groups are independent. The procedure to calculate this type of robust estimate begins by summing the scores within each respective cluster. The data set is thereupon collapsed so that there is only one observation per cluster or panel. A robust variance estimator is then determined in the same manner as in the noncluster case, except n is now the number of clusters and u consists of cluster sums. Refer to Table 3.11. A complete discussion of robust panel estimators is found in Hilbe (2009a) and Hardin and Hilbe (2013a).

The **medpar** data provide the hospital provider code with each observation. Called provnum, it is entered as an option to obtain the modified sandwich variance estimator. Unlike scaling and variance multipliers, robust estimators may be used with any maximum likelihood algorithm, not only GLM-based algorithms:

```
POISSON: CLUSTERING BY PROVIDER
glm los hmo white i.type, fam(poi) cluster(provnum)nolog

Generalized linear models                    No. of obs      =      1495
Optimization     : ML                        Residual df     =      1490
```

```
                                       Scale parameter =         1
Deviance       =  8142.666001          (1/df) Deviance = 5.464877
Pearson        =  9327.983215          (1/df) Pearson  = 6.260391
                                       AIC             = 9.276131
Log pseudolikelihood = -6928.907786    BIC             = -2749.057
                          (Std. Err. adjusted for 54 clusters in provnum)
-----------------------------------------------------------------------
             |             Robust
       los   |     Coef.   Std. Err.      z    P>|z|    [95% Conf. Interval]
-------------+---------------------------------------------------------
       hmo   | -.0715493    .0527299   -1.36   0.175   -.1748979    .0317993
     white   |  -.153871    .0729999   -2.11   0.035   -.2969482   -.0107939
      type   |
         2   |  .2216518    .0609139    3.64   0.000    .1022626    .3410409
         3   |  .7094767     .202999    3.49   0.000     .311606    1.107347
     _cons   |  2.332933    .0669193   34.86   0.000    2.201774    2.464093
-----------------------------------------------------------------------
```

Standard errors are produced by adjusting for the clustering effect on providers – that is, we suppose that the relationship between length of stay (*los*) and predictors is more highly correlated within a provider than between providers. This is a reasonable supposition, but in this case it appears that there is only a minimal clustering effect caused by providers. There is overdispersion, though – look at the unadjusted standard errors again. The empirical (robust) standard errors are double to triple the size of the model standard errors. Overdispersion results in the deflation of standard errors, which means that the denominator of coef / se is smaller and the value of z larger. As a result, overdispersion in general makes it appear that a predictor is significant (or contributes more to the understanding of the response) when in fact it is not. Smaller values of z are preferred, with a z of -1.96 being the two-sided .05 criterion for z (normal) significance based on the cumulative standard normal distribution:

```
STATA CODE                          R CODE
. di normal(-1.96)*2              > pnorm(-1.96)*2
.04999579                         [1] 0.04999579

. glm los hmo white i.type, fam(poi)  nolog nohead
-----------------------------------------------------------------------
             |              OIM
       los   |     Coef.   Std. Err.      z    P>|z|    [95% Conf. Interval]
-------------+---------------------------------------------------------
       hmo   | -.0715493     .023944   -2.99   0.003   -.1184786     -.02462
```

```
white |    -.153871    .0274128    -5.61   0.000    -.2075991    -.100143
 type |
    2 |    .2216518    .0210519    10.53   0.000     .1803908    .2629127
    3 |    .7094767     .026136    27.15   0.000     .6582512    .7607022
_cons |    2.332933    .0272082    85.74   0.000     2.279606     2.38626
-----------------------------------------------------------------------
> summary(poi1 < - glm(los ~ hmo+white+factor(type), family=poisson,
data=medpar))
```

Again we find overdispersion, but because there is little difference in SEs when comparing robust to robust clustered by *provnum*, we may conclude that there is no clustering effect based on hospital. Which hospital a patient is at makes no difference in their length of stay. The data can be *pooled*, meaning we do not need cluster (provnum) in the model. Note again that all summary statistics are the same as in the unadjusted model.

Statisticians frequently use empirical standard errors with Poisson regression as a catch-all adjustment for extradispersion, or any type of excess correlation in the data. Given the fact that the majority of Poisson models are extradispersed, many analysts simply employ empirical standard errors with all Poisson models. If the model is in fact equidispersed and the data are well fit using a Poisson model, the model and empirical standard errors will be nearly identical. If the model is over- or underdispersed, using empirical standard errors will provide the analyst with more accurate information regarding the significance of the predictors in explaining the count response. In addition, the exponentiated coefficients of a Poisson model with empirical standard errors can be referred to as rate ratios.

TABLE 3.13. R: Bootstrap Standard Errors

```
library(COUNT); library(boot); data(medpar)
poi <- glm(los ~ hmo + white + factor(type), family=poisson, data=medpar)
summary(poi)
t <- function (x, i) {
xx <- x[i,]
 bsglm <- glm( los ~ hmo + white + factor(type), family=poisson, data=medpar)
 return(sqrt(diag(vcov(bsglm))))
 }
bse <- boot(medpar, t, R=1000)
sqrt(diag(vcov(poi))); apply(bse$t,2, mean)
```

Consider a 2 × 2 table using the 1912 *Titanic* survival data:

```
STATA CODE
. use titanic
. gen died = survived
. recode died 1=0 0=1
. tab died age
                     0          1
          | Age (Child vs Adult)
    died |    child     adults |    Total
---------+--------------------+---------
       0 |       57        442 |      499
       1 |       52        765 |      817
---------+--------------------+---------
   Total |      109      1,207 |    1,316
```

The risk or probability of death for adults is

```
. di 765/1207
.63380282
```

The risk of death for children is

```
. di 52/109
.47706422
```

The risk or probability of death as an adult passenger on the *Titanic* compared with that for children is then

```
. di (765/1207) / (52/109)
1.3285482
. glm died age, fam(poi) nolog nohead vce(robust) ef
------------------------------------------------------------------------
         |              Robust
    died |      IRR   Std. Err.      z    P>|z|     [95% Conf. Interval]
---------+--------------------------------------------------------------
     age | 1.328548   .1364152    2.77   0.006    1.086366     1.62472
   _cons |  .4770642   .0478591   -7.38   0.000     .3919082    .5807235
------------------------------------------------------------------------
```

I do not show the calculations here, but the standard error for the risk of death based on *age* for those in the *Titanic* disaster is .134. Regardless of whether we use model or empirical standard errors with a Poisson regression, though (or negative binomial), the logic of the table analysis given here is the basis for interpreting coefficients in terms of probabilities. For example, here we may conclude that the likelihood or probability of death for adults is some 33% greater than it is for children. See Hilbe (2011) for a full estimation of risk ratio and risk difference.

3.4.4 Bootstrapped Standard Errors

The final method we discuss for adjusting count model standard errors in the face of overdispersion is the method known as the bootstrap (see Table 3.13 for R code). Nonparametric bootstrapping makes no assumptions about the underlying distribution of the model. Standard errors are calculated based on the data at hand. Samples are repeatedly taken from the data (with replacement), with each sample providing model estimates. The collection of vector estimates for all samples is used to calculate a variance matrix from which reported standard errors are calculated and used as the basis for determining confidence intervals. Such confidence intervals can be constructed from percentiles in the collection of point estimates, or from large sample theory arguments. The example that follows uses 1000 samples of 1495; each sample provides an estimated coefficient vector from which standard errors are calculated. The number of samples may be changed.

```
. bootstrap, reps(1000): glm los hmo white type2 type3, fam(poi)

Generalized linear models                No. of obs      =       1495
Optimization      : ML                   Residual df     =       1490
                                         Scale parameter =          1
Deviance       =   8142.666001          (1/df) Deviance =   5.464877
Pearson        =   9327.983215          (1/df) Pearson  =   6.260391
                                         AIC             =   9.276131
Log likelihood =  -6928.907786           BIC             =  -2749.057
-----------------------------------------------------------------------
             |   Observed   Bootstrap                  Normal-based
         los |      Coef.   Std. Err.     z    P>|z|   [95% Conf. Interval]
-------------+---------------------------------------------------------
         hmo |  -.0715493   .0527641   -1.36   0.175   -.174965    .0318664
       white |   -.153871   .0815228   -1.89   0.059  -.3136529    .0059108
       type2 |   .2216518   .0536495    4.13   0.000   .1165006    .3268029
       type3 |   .7094767   .1178019    6.02   0.000   .4785893    .9403641
       _cons |   2.332933    .077869   29.96   0.000   2.180313    2.485554
-----------------------------------------------------------------------
```

Standard errors indicate that *hmo* and *white* are problematic. Bootstrapping may be done on parts of a model other than standard errors. With respect to standard errors, though, bootstrapping has become a popular way of attempting to discover optimal standard errors for model coefficients. In fact, bootstrapping and using robust or empirical standard errors is an excellent way to determine whether a model is overdispersed, or extradispersed in general:

> *If the values of bootstrapped or robust standard errors differ substantially from model standard errors, this is evidence that the count model is extradispersed. Use the bootstrapped or robust standard errors for reporting your model, but check for reasons why the data are overdispersed and identify an appropriate model to estimate parameters.*

A note should be made regarding *profile likelihood confidence intervals*. R's **glm** function uses them as the default for calculating confidence intervals. Stata has a user-authored **logprof** command for constructing profile confidence intervals for logistic regression models only. Recall that I have said that using the *likelihood ratio test* (discussed in Section 4.2) for determining the inclusion or exclusion of predictors in a model is preferred over the standard Wald method, which is another way of saying regular predictor *p*-values. The profile likelihood method is based on the likelihood ratio test but is inverted to move from the *p*-value to the confidence interval rather than have a confidence interval and then seek to determine whether it includes 0. Except in R, the method is rarely used, and it has limited software support. But it is an attractive method of presenting confidence intervals, and I suspect that it will be more commonly used in the next decade. I refer you to Hilbe and Robinson (2013) for a more detailed examination of profile likelihood, together with R code for its implementation.

3.5 SUMMARY

Overdisperson is likely the foremost problem with count data. Overdispersion also occurs with grouped logistic regression, as well as grouped binomial models as a class. It also is a concern for ordered, partially ordered, and unordered or multinomial regression models as well. But when a statistician mentions that he or she is working with overdispersed data, the usual supposition is that they are count data. This is because most count models are (Poisson) overdispersed (i.e., the variance of the count response variable is greater than the mean).

In this chapter, we looked at what overdispersion means and the difference between apparent and real overdispersion. We also discussed various statistical tests that can be given a Poisson model to determine whether it is overdispersed or underdispersed. I also provided an overview and code that can be adapted to the data for determining the relationship between the count values that are in fact observed and those that are predicted on the basis of a statistical model. Ideally they should be close. A Chi2 test can be used to determine how well they fit, but such a test is not as feasible when many of the counts are skewed far to the right.

Finally, we discusssed strategies for handling overdispersion – in particular light overdispersion. A surprisingly good way to adjust for extradispersion is to scale the standard errors by the square root of the Pearson dispersion, which is known as the *quasipoisson* "family" in R. This is popular among R users. Most specialists in count models, however, espouse using robust or sandwich adjusted standard errors when modeling count data – even as a default. If there is no correlation in the data to be adjusted, the SEs of the sandwich estimator will be the same as model SEs. Some statisticians prefer to use bootstrapped SEs in place of sandwich SEs, but bootstrapping takes longer and the results are typically the same as for the robust or sandwich approaches.

## CHAPTER 4

Assessment of Fit

SOME POINTS OF DISCUSSION

- How are residuals analyzed for count data? Are they different from those for logistic models?
- What is a boundary likelihood ratio test? How does it differ from a regular likelihood ratio test?
- What are information criterion tests? How does one select the most appropriate one to use?
- When is it important to use a validation sample? How big should it be?

4.1 ANALYSIS OF RESIDUAL STATISTICS

When modeling, using either full Newton–Raphson maximum likelihood or IRLS, it is simple to calculate the linear predictor as

$$x\beta = \eta = \alpha + \beta_1 + \beta_1 + \cdots + \beta_n$$

or

$$x\beta = \eta = \beta_0 + \beta_1 + \beta_2 + \cdots + \beta_n$$

where α or β_0 is the intercept term, defined as the value of the linear predictor of model observations when the value of each predictor is 0. Remember, however, for continuous or count variables that do not have a 0 value (e.g.,

TABLE 4.1. Count Model Residual Formulae: Poisson Residuals		
Pearson		$R^p = (y - \mu)/\text{sqrt}(V)$ (4.1)
	Poisson	$(y - \mu)/\sqrt{\mu}$
Deviance		$R^d = \text{sgn}(y - \mu)*\text{sqrt(deviance)}$ (4.2)
	Poisson	$R^d = \text{sign}(y - \mu)\sqrt{2\left\{y \log\left(\frac{y}{\mu}\right) - (y - \mu)\right\}}$
Standardized		Divide residual by sqrt(1−hat), which aims (4.3) to make the variance constant. hat = stdp^2 * V
	Poisson	$\dfrac{R^p}{\sqrt{1 - h}}$
Studentized		Divide standardized residual by scale, ϕ. (4.4)
	Poisson	$\dfrac{R^p}{\sqrt{\phi}}$
Standardized-Studentized		Divide by both standardized and (4.5) studentized adjustments; e.g., R^p: $(y - \mu)/\{\phi V(\mu)*\text{sqrt}(1 - h)\}$
	Poisson	$\dfrac{R^p}{\sqrt{\phi(1 - h)}}$

an *age* variable ranging from 25 to 64), the intercept will not equal the linear predictor. In any case, each observation in the model has a linear predictor value, $x\beta$ or η. For members of the GLM family, including Poisson and negative binomial regressions, a link function converts a linear predictor to a fitted or predicted value. We have previously seen that for Poisson models as well as for the traditional negative binomial, $\eta = \ln(\mu)$, so that $\mu = \exp(\eta)$ or, for full maximum likelihood models, $\mu = \exp(x\beta)$. Simply put, exponentiating the linear predictor produces the fitted, predicted, or expected value of the count response. The linear predictor and fit are essential components of all residuals.

The basic or raw residual is defined as the difference between the observed response and the predicted or fitted response. When y is used to identify the response, $\hat{y}$ or μ is commonly used to characterize the fit. Hence

$$\text{Raw residual} = y - \hat{y} \text{ or } y - \mu \text{ or } y - E(y)$$

Other standard residuals used in the analysis of count response models include those in Table 4.1.

In the preceding formulae, we indicated the model variance function as V, the hat matrix diagonal as hat or h, and the standard error of the prediction as *stdp*. A scale value, ϕ, is user defined and is employed based on the type of data being modeled.

We mentioned earlier that the Anscombe residual (Anscombe 1953) has values close to those of the standardized deviance. There are times, however, when this is not the case, and the Anscombe residual performs better than R^d. Anscombe residuals attempt to normalize the residual so that heterogeneity in the data, as well as outliers, become easily identifiable.

Anscombe residuals use the model variance function. The variance functions for the three primary count models, as well as the PIG model, are

Poisson $V = \mu$

Geometric $V = \mu(1 + \mu)$

NB2 $V = \mu + \alpha\mu^2$ or $\mu(1 + \alpha\mu)$ # R's **glm.nb** uses $\mu + (\mu^2)/\theta$, with $\theta = 1/\alpha$

PIG $V = \mu + \alpha\mu^3$ # some apps use $\mu + (\mu^3)/\alpha$

The geometric distribution is the negative binomial with the scale or dispersion parameter, α, equal to 1. It is rarely estimated as a geometric model since true geometric data would have a dispersion parameter value of 1 when modeled using a negative binomial regression.

Anscombe defined the residual that later became known under his name as

$$R^A = \frac{A\,(y) - A\,(\mu)}{A'\,(\mu)\,\sqrt{V}} \tag{4.6}$$

here $A\,(\bullet) = \int V^{1/3}$.

The calculated Anscombe residuals for the Poisson model are

$$\text{Poisson: } 3(y^{2/3} - \mu^{2/3})/(2\mu^{1/6}) \tag{4.7}$$

and for the negative binomial

$$\frac{\left\{\frac{3}{\alpha}\left[(1 + \alpha y)^{2/3} - (1 + \alpha\mu)^{2/3}\right] + 3\left(y^{2/3} - \mu^{2/3}\right)\right\}}{2\left(\alpha\mu^2 + \mu\right)^{1/6}} \tag{4.8}$$

The Anscombe negative binomial has also been calculated in terms of the hypergeometrix2F1 function. See Hilbe (1993c) and Hardin and Hilbe (2012) for a complete discussion.

$$y^{2/3}H\,(2/3, 1/3, 5/3, y/\alpha) - \mu^{2/3}H\,(2/3, 1/3, 5/3, \mu/\alpha) \tag{4.9}$$

$$= 2/3\, B\,(2/3, 2/3)\{y - B_1(2/3, 2/3, \mu/\alpha)\} \tag{4.10}$$

where H is the hypergeometrix2F1 function, B is the *beta* function, and B_I is the incomplete *beta* function. Hilbe (1994a) and Hardin and Hilbe (2012) show that the two-term *beta* function has the constant value of 2.05339. α is the negative binomial dispersion parameter.

Analysts generally prefer to graph the standardized Pearson or Anscombe residuals by μ. The default residual in R is the deviance. Pearson residuals may be obtained after a Poisson model of the **rwm1984** data as follows. The model is named *pexp*.

```
> summary(pexp <- glm(docvis ~ outwork + cage, family=poisson,
data=rwm1984))
> presid <- residuals(pexp, type="pearson")
```

The Pearson Chi2 statistic is calculated as the sum of the squared Pearson residuals. Put in one line, we have

```
> pchi2 <- sum(residuals(pexp, type="pearson")^2)  # Pearson Chi2
```

What are we looking for in graphing residuals for count models? We look for two things:

- evidence of poor fit, and
- nonrandom patterns.

We can evaluate these two goals by graphing or plotting the standardized deviance or Anscombe residual by using

- the predicted or fitted value, μ.

Patterns usually indicate overdispersion and/or misspecification. An alternative count model may be required:

- all model predictors.

Patterns typically mean that observations are not independent, or perhaps proneness exists in the data. It may also indicate that one or more predictors need to be converted to another scale (e.g., squared or logged). Finally, a smoother may need to be added to the model terms:

- predictors we found not to contribute to the model.

Try including the excluded predictors, but they may need rescaling or are part of an interaction term:

- a time variable (for longitudinal and time series count models).

Independence is violated. Use a GEE or mixture model:

```
> summary(rwm <- glm(docvis ~ outwork + age, family=poisson,
data=rwm1984))
                              < output not displayed>
> P__disp(rwm)
 Pearson Chi2 =   43909.62
 Dispersion   =   11.34322

> mu <- predict(rwm)
> grd <- par(mfrow = c(2,2))
> plot(x=mu, y= rwm$docvis, main = "Response residuals")
> plot(x=mu, y= presid, main = "Pearson residuals")
<<Figures not displayed>>
```

4.2 LIKELIHOOD RATIO TEST

We mentioned the likelihood ratio test earlier in the book. It is a key test for assessing the worth of nested models (i.e., where a model with fewer predictors is compared with the same model with more predictors). The test evaluates whether the predictors withdrawn from a model should in fact have been retained. Likelihood ratio tests also are used to compare different models if one is a subset or reduced version of another. For example, one may use a likelihood ratio test to determine whether data should be modeled using a Poisson or negative binomial regression. This is a crude way of expressing the relationship, but it is how most analysts think of it. We use what has been termed a *boundary likelihood ratio test* for testing the Poisson model versus the negative binomial model (see Hilbe 2007a, 2011).

4.2.1 Standard Likelihood Ratio Test

The traditional likelihood ratio test is defined as

$$\text{LR} = -2(\mathcal{L}_R - \mathcal{L}_F) \tag{4.11}$$

where $\mathcal{L}_F$ is the log-likelihood for a full or more complete model and $\mathcal{L}_R$ is the log-likelihood for a reduced model. An example will help.

We use the German health data, with both *outwork* and *age* as predictors:

```
. use rwm1984,clear
. qui glm docvis outwork age, fam(poi)   // full model
. est store A
```

```
. qui glm docvis outwork, fam(poi)        // reduced model, drop age
. est store B
. lrtest A B

Likelihood-ratio test                      LR chi2(1)  =     715.98
(Assumption: B nested in A)                Prob > chi2 =     0.0000
```

The test confirms that *age* is a significant predictor in the model and should be retained. The likelihood ratio **drop1** test is a useful likelihood ratio test in which one predictor at a time is dropped and the models are checked in turn for comparative goodness-of-fit. In Stata, a user-developed command (Wang 2000) called **lrdrop1** can be used following **logit**, **logistic**, and **poisson** to find the predictors that together best fit the model. **lrdrop1** can rather easily be adapted for use with other regression models. It is particularly useful when a model has quite a few potential predictors. **lrdrop1** must be installed using the command ".**ssc install lrdrop1**" in order to use it the first time:

```
STATA OUTPUT
. qui poisson docvis outwork age
. lrdrop1
Likelihood Ratio Tests: drop 1 term
poisson regression
number of obs = 3874
---------------------------------------------------------------------
  docvis    Df     Chi2    P>Chi2    -2*log ll   Res. Df    AIC
---------------------------------------------------------------------
Original Model                        31272.78    3871    31278.78
-outwork     1    464.77    0.0000    31737.55    3870    31741.55
   -age      1    715.98    0.0000    31988.76    3870    31992.76
---------------------------------------------------------------------
Terms dropped one at a time in turn.
```

R users are familiar with both **lrtest** and especially **drop1** (see Table 4.2). The latter uses the deviance statistic rather than the log-likelihood as the basis for model comparison. Recall that the deviance is itself derived as a variety of likelihood ratio test and is defined as 2{LLs – LLm}, where LLm is the log-likelihood of the model and LLs is the saturated log-likelihood. For LLs, the response term, y, is substituted for each μ in the log-likelihood equation.

The p-value of 2.2e-16 indicates that *outwork* and *age* are significant predictors in the model.

The **drop1** test found in R is a likelihood ratio test of each nested predictor in a model. Deviance and AIC statistics are produced for each alternative.

TABLE 4.2. R: Likelihood Ratio Test

```
===========================================================
library(COUNT); library(lmtest); data(rwm5yr)
rwm1984 <- subset(rwm5yr, year==1984)
poi1 <- glm(docvis ~ outwork + age, family=poisson, data=rwm1984)
poi1a <- glm(docvis ~ outwork, family=poisson, data=rwm1984)
lrtest(poi1, poi1a)
drop1(poi1, test="Chisq")
===========================================================

R OUTPUT
Likelihood ratio test

Model 1: docvis ~ outwork + age
Model 2: docvis ~ outwork
  #Df LogLik Df  Chisq Pr(>Chisq)
1   3 -15636
2   2 -15994 -1 715.98  < 2.2e-16 ***
---
Signif. codes:  0 `***' 0.001 `**' 0.01 `*' 0.05 `.' 0.1 ` ' 1

> drop1(poi1,test="Chisq")
Single term deletions

Model:
docvis ~ outwork + age
        Df Deviance   AIC    LRT  Pr(>Chi)
<none>        24190 31279
outwork  1    24655 31742 464.77 < 2.2e-16 ***
age      1    24906 31993 715.98 < 2.2e-16 ***
---
Signif. codes:  0 `***' 0.001 `**' 0.01 `*' 0.05 `.' 0.1 ` ' 1
```

The test is used by many analysts as a substitute for the standard likelihood ratio test.

For the purposes of modeling counts, the foremost use of the test is when it is used as a boundary test.

4.2.2 Boundary Likelihood Ratio Test

The boundary likelihood ratio (BLR) test is a test used on negative binomial models to determine whether the value of the dispersion parameter, α, is significantly different from 0. We will discuss the negative binomial model in the next chapter, so I do not want to go into detail here.

It is important, though, to remember that the BLR test has a lower limiting case for the value of α, which is what is being tested. Given that the standard parameterization of the negative binomial variance function is $\mu + \alpha\mu^2$, when $\alpha = 0$, the variance reduces to μ. The Poisson variance function is μ, the same as its mean value.

The BLR equation is given as

$$-2(\mathcal{L}_P - \mathcal{L}_{\mathrm{NB}}) \tag{4.12}$$

with $\mathcal{L}$ symbolizing the log-likelihood function. The resulting value is measured by an upper tail Chi2 distribution with one (1) degree of freedom. Since the distribution being tested can go no lower than 0, that is the boundary. Only one half of the full distribution is used. Therefore the Chi2 test is divided by 2. For a standard Chi2 test with no boundary, with one degree of freedom, a BLR test statistic of 3.84 is at the .05 significance level:

```
. di chi2tail(1,3.84)
.05004352
```

Dividing by 2 means that a value over 2.705 is not significant:

```
. di chi2tail(1,2.705)/2
.05001704
```

In R, the **pchisq** function must be used, giving the following result:

```
> pchisq(2.705,1, lower.tail=FALSE)/2
[1] 0.05001704
```

As a key to remember, given the value of -2 times the difference in log-likelihood values, if the difference in Poisson and negative binomial log-likelihoods is less than 1.352, the model is Poisson; if it is greater, the data need to be modeled other than Poisson – presumably negative binomial. We will discover that there are quite a few count models that are nested within another, for which a likelihood ratio test is appropriate (e.g., zero-truncated Poisson and zero-truncated negative binomial, censored Poisson and censored negative binomial).

4.3 MODEL SELECTION CRITERIA

4.3.1 Akaike Information Criterion

The Akaike Information Criterion (AIC) was the creation of Japanese statistician Hirotsugu Akaike (1973). However, the AIC statistic did not begin to have widespread use until the twenty-first century. Actually, Akaike later developed what he considered to be a superior statistic, called the ABIC, but it has not gained acceptance by the greater research community. The AIC is the most commonly used general fit statistic.

The AIC statistic is generally found in two forms, the traditional version

$$\text{AIC} = -2\mathcal{L} + 2k = -2\,(\mathcal{L} - k) \tag{4.13}$$

and the version with the main AIC terms divided by n, the number of observations in the model

$$\text{AIC} = \frac{-2\mathcal{L} + 2k}{n} = -2\,(\mathcal{L} - k)/n \tag{4.14}$$

where $\mathcal{L}$ is the model log-likelihood, k is the number of predictors, including the intercept, and n the number of observations in the model. $2k$ is referred to as a *penalty* term, which adjusts for the dimension of the model. Given that adding more parameters to a model makes the data more likely, as we increase the number of predictors, $-2\mathcal{L}$ becomes smaller. The penalty, $2k$, is added to the log-likelihood to adjust for this possible bias.

Larger n also affects the $-2\mathcal{L}$ statistic. Equation (4.14) divides the main terms in the AIC by n, which makes the AIC/n a per-observation contribution to the adjusted $-2\mathcal{L}$. All other conditions being equal, when comparing models, the better-fitted model has a smaller AIC statistic. There is no difference in this respect between equations (4.13) and (4.14).

R (default package) and Stata adopt equation (4.13) as the primary definition of AIC, assuming that both equations (4.13) and (4.14) have the same inherent information (i.e., both provide equal assessments regarding comparative model fit). However, this is not necessarily the case, especially when the statistic is used for assessing the fit of models whose observations are not independent, or when the statistic is used to compare models of different sizes, which can happen with nested models where one or more of the predictors have missing values. The AIC is also used by some analysts to compare models of different sample size. Although the AIC test can be used on nonnested models, the response variable of models being compared must be the same.

The analyst must be cautioned to be aware of which meaning or definition of AIC is being used in reporting a study. In general, low values of AIC compared with the number of model observations indicate that equation (4.14) is being used; large values indicate equation (4.13).

Other AIC statistics have been used in research. The most popular – other than the two primary versions given here – is the *finite sample AIC*, which may be defined as

$$\mathrm{AIC_{FS}} = -2\{\mathcal{L} - k - k(k+1)/(n-k-1)\}/n \qquad (4.15)$$

or

$$\mathrm{AIC_{FS}} = \mathrm{AIC} + \frac{2k\,(k+1)}{n-k-1} \qquad (4.16)$$

where k is the number of parameters in the model. This statistic has also been referred to as $\mathrm{AIC_C}$ for AIC-corrected, but I will use that designation for equation (4.18). Note that $\mathrm{AIC_{FS}}$ has a greater penalty for additional parameters compared with the standard AIC statistic. Note also that $\mathrm{AIC_{FS}} \approx \mathrm{AIC}$ for models with large numbers of observations. Hurvich and Tsai (1989) first developed the finite-sample AIC for use with time series and autocorrelated data; however, others have considered it preferable to the AIC for models with noncorrelated data as well, particularly for models with many parameters, and/or for models with comparatively few observations. When used for longitudinal and time series data, the n in equation (4.16) is summed across panels, $\Sigma_i^N n_i$. A variety of finite-sample AIC statistics have been designed since 1989.

A *Consistent Akaike Information Criterion* (CAIC) test and *Corrected AIC* have also enjoyed some popularity. Defined as

$$\mathrm{CAIC} = -2\mathcal{L} + k^* \log(n) \qquad (4.17)$$

and

$$\mathrm{AICC} = -2\mathcal{L} + \frac{2\,kn}{n-k-1} \qquad (4.18)$$

these tests are sometimes found in commercial statistical software, such as the AICC in SAS. It is wise to be aware of their existence and meaning since you will find them being used in journal articles and books on modeling.

Recently, the AIC statistic and alternatives have been the focus of considerable journal article research, particularly with respect to correlated data. Hin and Wang (2008) and Barnett et al. (2010) even demonstrated that the AIC is superior to the QIC statistic, which is an AIC-type fit statistic designed for generalized estimating equation (GEE) longitudinal models by Pan (2001) and revised by Hardin and Hilbe (2002). The QIC statistic has itself been revised in light of these objections, most recently by Hardin and Hilbe (2013) and Shults and Hilbe (2014). These applications are for longitudinal data, or data structured in panels.

The traditional AIC statistic can be enhanced to provide a superior general comparative-fit test for maximum likelihood models. Simulation tests performed by Hardin and Hilbe (2013) appear to confirm that the following version of AIC, which we may tentatively call AIC$_H$, produces a test statistic that is aimed at optimally selecting the best fitted among competing maximum likelihood models. The statistic is robust to violations of MLE assumptions, such as the requirement of the independence of observations. There are three adjusters in the statistic – the number of observations in the model, number of parameters (coefficients), and number of distributional parameters – for example, the mean or location and scale parameters. In special circumstances, a scale parameter can become a dispersion parameter:

$$\text{AIC}_H = -2\mathcal{L} + \frac{4(p^2 - pk - 2p)(p + k + 1)(p + k + 2)}{n - p - k - 2} \quad (4.19)$$

with

$n = $ number of observations in the model;

$p = $ number of β slopes in the model;

$k = $ number of model location and scale parameters.

This test statistic is still a work in progress but appears to correctly select the best fitted among competing maximum likelihood estimated models. I will show the results of using it together with other more traditional versions in subsequent example model output.

Let's return to the traditional AIC statistic currently provided in the majority of study results. How do we decide whether one value of the AIC is significantly superior to another; that is, whether there is a statistically significant difference between two values of the AIC? Hilbe (2009a) devised a table based on simulation studies that can assist in deciding whether the difference between two AIC statistic values is significant. Table 4.3 is taken from that source and is based on equation (4.13) (i.e., the AIC without division by n).

TABLE 4.3. AIC Significance Levels	
Difference between Models A and B	Result if A < B
>0.0 & <= 2.5	No difference in models
>2.5 & <= 6.0	Prefer A if n > 256
>6.0 & <= 9.0	Prefer A if n > 64
>9.0	Prefer A

Table 4.3 is a general guide only and should be used as an approximation. An example of AIC and BIC statistics will be given in the following subsection. If a difference does not fit into one of the preceding categories, then there is no preference between the alternative models.

4.3.2 Bayesian Information Criterion

The *Bayesian Information Criterion* (BIC) was first formulated by Gideon Schwarz in 1978. It is the second foremost contemporary comparative-fit statistic for maximum likelihood models found in both commercial statistical packages and freeware, such as R. The procedure is at times known by different names (e.g., the Schwarz Criterion, or simply SC in SAS). It is formulated as

$$\text{BIC}_\text{L} = -2\mathcal{L} + k\log(n) \tag{4.20}$$

with k indicating the number of predictors, including the intercept, and n the number of observations in the model. The statistic gives a higher weight to the adjustment term, $k*\log(n)$, than does the AIC statistic, where only $2k$ is employed for adjusting -2 times the model log-likelihood function. It is on this account that many statisticians prefer the Schwarz BIC to the traditional AIC, but in general the AIC is used more frequently by analysts when testing and reporting research studies.

Another parameterization of the BIC, which is used in the Limdep software program, was developed by Hannan and Quinn (1979). It is given as

$$\text{BIC}_\text{HQ} = -2(\mathcal{L} - k*\ln(k))/n \tag{4.21}$$

TABLE 4.4. R: Version of Stata User Command `abic`

```
===============================================================
modelfit <- function(x) {
obs <- x$df.null + 1
aic <- x$aic
xvars <- x$rank
rdof <- x$df.residual
aic_n <- aic/obs
ll <- xvars - aic/2
bic_r <- x$deviance - (rdof * log(obs))
bic_l <- -2*ll + xvars * log(obs)
bic_qh <- -2*(ll - xvars * log(xvars))/obs
c(AICn=aic_n, AIC=aic, BICqh=bic_qh, BICl=bic_l)
}
modelfit(x) # substitute fitted model name for x
===============================================================

library(COUNT)
data(medpar)
mymodel <- glm(los ~ hmo + white + factor(type), family=poisson, data=medpar)
modelfit(mymodel)
  <not diisplayed>
```

which has results that approximate AIC/n in equation (4.14) for most models. If the values differ substantially, this is an indication that the model is poorly fit and is likely misspecified. This test is not definitive, though, since many poorly fit models have similar AIC/n and BIC_{HQ} values. Many statisticians, including the author, prefer the use of BIC_{HQ} to the Schwarz parameterization.

The Stata **abic** command and the R **modelfit** function (in the **COUNT** package) display the values of the most used parameterizations of the AIC and BIC statistics. The BIC statistic given in the lower-left corner of the output is BIC_{HQ}, whereas the lower-right BIC statistic is BIC_{SC}, or Schwarz BIC. The top-left AIC is equation (4.14) and the top right is equation (4.13). The way that the labeling reads is that the AIC by default is (4.13). Multiplying equation (4.13) by n yields equation (4.14). R's **modelfit.r** function is provided in Table 4.4. The AIC_H command was added during the writing of this book and is displayed in selected model output.

An example will perhaps help clarify things. Using the **medpar** data, we model and use the **abich** (**abic** + AIC_H) and **modelfit** statistics for Stata and R output, respectively:

```
. glm los hmo white type2 type3, fam(poi) nolog nohead
------------------------------------------------------------------------
             |               OIM
     los |      Coef.   Std. Err.      z    P>|z|     [95% Conf. Interval]
-------+----------------------------------------------------------------
     hmo |  -.0715493    .023944    -2.99   0.003    -.1184786    -.02462
   white |   -.153871   .0274128    -5.61   0.000    -.2075991   -.100143
   type2 |   .2216518   .0210519    10.53   0.000     .1803908   .2629127
   type3 |   .7094767    .026136    27.15   0.000     .6582512   .7607022
   _cons |   2.332933   .0272082    85.74   0.000     2.279606    2.38626
------------------------------------------------------------------------

. abich
AIC Statistic   =    9.276131          AIC*n       = 13867.815
BIC Statistic   =    9.280208          BIC(Stata)  = 13894.365
AICH Statistic  =    9.274205          AICH*n      = 13854.255
```

Stata's **estat ic** command is commonly used by Stata users to obtain AIC/BIC statistics after modeling maximum likelihood and GLM models:

```
. estat ic
------------------------------------------------------------------------
   Model |     Obs   ll(null)   ll(model)     df        AIC         BIC
-------+----------------------------------------------------------------
       . |    1495          .   -6928.908      5    13867.82    13894.36
------------------------------------------------------------------------
```

Note that in the Stata **glm** command output the BIC statistic displayed in the header statistics is commonly referred to as the *Raftery BIC*, authored in 1986 by Adrian Raftery of the University of Washington. It is based on the deviance function, not the log-likelihood as are other AIC and BIC statistics. The reason for this is Raftery's desire to implement an information criterion fit test for generalized linear models, which were at the time all based on the deviance, with the log-likelihood rarely even calculated. Raftery's BIC is seldom used at this time. I wrote Stata's initial **glm** command in late 1992, when the deviance was still used as the basic fit test in GLM softare, as well as the criterion for convergence. The Raftery BIC test statistic is defined as

$$BIC_R = D - (df)\ln(n) \qquad (4.22)$$

```
bic <- model$deviance - (model$df.residual * log(model$df.null +1))
```

The AIC and BIC statistics are now considered as basic fit tests and should normally accompany statistical model output. The traditional AIC and Schwarz BIC are assumed when a reference is made to AIC and BIC statistics. If you select another definition, be certain to mention it in a research report.

4.4 Setting up and Using a Validation Sample

Analysts commonly use a validation data set to help confirm the generalizability of the model they are developing to a greater population. This is particularly the situation when the data come from a sample of observations within a known population. This concept of extractability is inherent in the frequency interpretation of statistics, on which the models we are discussing are based.

Validation data generally come in two varieties: (1) as a sample of population data from which the model data derive but that are not included in the model and (2) a sample taken from the model itself. Percentages differ, but validation samples typically consist of about 20% of the observations from the model being examined – however, there is no hard rule as to the actual percentage used. If validation data are taken from the estimated model, the procedure is to model the validation data in exactly the same manner as in the original model. The validation model coefficients and standard errors should be similar to those produced by the original model. A *Hausman test* (Hausman 1978) on the equality of coefficients can be used to assess the statistical difference in the two groups of observations.

A given percentage of data may also be withheld from the estimation of the initial or primary model. Once the selected data have been modeled, the withheld validation data are subsequently modeled for comparison purposes. Using a test like Hausman's, the coefficients are tested for similarity. If it appears that the two sets of data come from the same population – that the data are not statistically different – then the two sets of data may be combined and remodeled for a final fitted model.

Both of these methods of constructing and implementing a validation sample help prevent an analyst from developing an overfitted model. Overfitted models are such that they cannot be used for classification, or for prediction of nonmodel data from a greater population. An overfitted model is so well fitted that it is applicable only to the data being modeled. Coefficient values and other summary statistics may not be extrapolated to other data. For most research situations, such a model is of little value. Unfortunately, though, many analysts attempt to fit a model as perfectly as possible so that overfitting does become a problem.

It is easy to overfit a model. When an analyst incorporates a large number of predictors into the model and then attempts to fine-tune it with highly specific transformations and statistical manipulations, it is likely that the resultant model is overfitted. Testing a model against validation data assists in minimizing this possibility and helps assure us of the stability of the parameter estimates.

Remember that even if we are confident that a model is well fitted, it is nevertheless dangerous to make predictions outside the sample space of the model. If the modeled data are truly representative of the population on which the modeled data is a sample, it is safer to make out-of-sample predictions, but it's not wise to make them too far from the model's sample space.

How far is too far? Let experience and common sense be a guide. Statistics is in many respects an art, not simply a recipe to be followed.

4.5 SUMMARY AND AN OVERVIEW OF THE MODELING PROCESS

Chapter 2 and 3 are the longest of the book, with Chapter 3 the longer. The reason for such lengthy chapters is the fact that the Poisson model rests at the base of all other count models. Chapter 3 deals with the foremost problem that must be addressed when modeling counts – overdispersion. We described what overdispersion entails, how to test for it, how to handle it, and how to assess the fit of count models in general. In particular, the Poisson model is used for the example regression model for these tests. For the most part, tests like the likelihood ratio test, the AIC and BIC statistics, and so forth are applicable to many of the models we discuss in the remainder of the book. However, keep in mind that AIC/BIC type statistics assume that observations are independent in a model. This is not always the case. They are fairly robust to violations of these assumptions, though. Many analysts still use the tests, even for random-effects models and other models that are plainly correlated.

When used in these circumstances, the information criterion tests (AIC, BIC, etc.) are likely valid indicators of which model is better fit, but only if there is a wide separation in AIC/BIC values between the models being compared. I caution you not to rely solely on them for assessing comparative model fit when the data are clearly correlated.

When modeling, it is important to mention that analysts should not focus on predictor p-values. Rather, emphasis should be given to confidence intervals. Moreover, unless the distributional assumptions of the model are met in full, it is preferred to employ robust or sandwich-based confidence intervals. Profile and bootstrapped confidence intervals are also preferred. If the lower and upper 95% confidence limits of a coefficient exclude 0, or if a risk ratio excludes 1, we may be 95% confident that the true value of the coefficient or risk ratio lies somewhere within the confidence limits.

A predictor p-value relates to the value of the predictor z-statistic, which itself is the ratio of the coefficient to the standard error. It is not the probability that the predictor fits the model, or of how influential a predictor is in the model. Unfortunately many analysts interpret a p-value in this manner. A more full discussion of p-values and confidence intervals takes us beyond the scope of this guidebook. I refer you to Vickers (2010) and Ellis (2010) for more details. It is advised, though, to give preference to confidence intervals when considering a predictor's worth to a model.

At this point, we start our examination of models that extend the base Poisson model. Most are designed to adjust for overdispersion in general, whereas others adjust for a specific distributional abnormality. Several models we discuss are also capable of adjusting for underdispersion.

First, however, I would like to provide a summary of what we have thus far discussed regarding the modeling process.

4.5.1 Summary of What We Have Thus Far Discussed

As a general approach to modeling count data, let's review a bit how to start the modeling process. I assume you have looked carefully at the data. For binary variables, check to see whether one of the levels has only a small percentage of the observations or there are missing values. For categorical variables, check to see whether one or more levels are markedly unbalanced, with too many or too few observations. The more balanced the predictor, the better the model that can be developed. For continuous predictors, check the range – the minimum and maximum values – and whether the mean and median are nearly the same. If they are, the variable may be normally distributed (i.e., bell shaped). If a continuous variable such as age begins at 40 and ends at 65, it is likely best to center it. This means that the variable, as a predictor in a model, will be interpreted differently from normal. Check also for outliers and missing values. This is all preliminary to actually modeling the data.

For count models such as Poisson or negative binomial regression, I suggest using robust standard errors as a default method. If there is no excessive correlation in the data, the robust standard errors will reduce to the values of model standard errors. If there is correlation in the data, which is typically the case, then the robust standard errors help adjust for the extradispersion. Bootstrapping does the same thing but takes longer, and the results are nearly identical.

If overdispersion is not adjusted by simple post hoc standard error adjustment, the majority of analysts turn to negative binomial regression. Of course, if one's study data are structured such that no zero (0) counts are allowed, or if there are far more zero (0) counts than allowed by the distributional assumptions of the Poisson model, then using a zero-truncated or zero-inflated Poisson model might be the preferred approach. We'll discuss this scenario later. On the other hand, if an analyst is not sure of the cause of model overdispersion, a good procedure is to model the data using a negative binomial model, testing it using a boundary likelihood ratio test. There are other options as well, but this is a good start.

To reiterate, we next address a model that can be used for generic overdispersion – one for which we may not know the source of extra correlation in the data. Of course, for the example **rwm1984** data we have been modeling, I strongly suspect that the excessive zero counts found in the data result in overdispersion. For the **medpar** data, the response term, *los*, has no possibility of having zero counts. That situation also gives rise to overdispersion – but for a different reason. Remember also that a model may have multiple sources of overdispersion. Adjusting for one source will seldom adjust for other sources. For now, however, we will overlook these facts and will assume that we have no idea why our example models are overdispersed.

## CHAPTER 5

Negative Binomial Regression

SOME POINTS OF DISCUSSION

- How does the negative binomial model differ from the Poisson model?
- What is the difference between NB2 and NB1 models? Which is the traditional version? Why?
- What are the assumptions of the negative binomial model?
- What is the difference between the dispersion statistic and parameter? How do they relate?
- What is NB-P, and how can it help select the best fit between NB1 and NB2 models?
- Can the dispersion parameter also be parameterized – a heterogeneous negative binomial?

5.1 VARIETIES OF NEGATIVE BINOMIAL MODELS

I have mentioned several times thus far in our discussion that the standard or traditional understanding of negative binomial regression is as a Poisson-gamma distribution mixture model with a mean of μ and a variance of $\mu + \alpha\mu^2$ or $\mu(1 + \alpha\mu)$. The mean is understood in the same manner as the Poisson mean, but the variance has a much wider scope than is allowed by the Poisson distribution.

The traditional parameterization of the negative binomial, as given in the previous paragraph, is also known as the NB2 negative binomial model, based on the value of the exponent in its second term. The NB2 model is also referred to as a *quadratic* negative binomial. An NB1 model has also been formulated for which the second term's exponent has a value of 1. It is a *linear* negative binomial, with a mean of μ and a variance of $\mu + \alpha\mu$. Both NB1 and NB2 use a maximum likelihood algorithm for estimating parameters, but NB2 may also be estimated using an IRLS algorithm within the scope of generalized linear models.

The negative binomial is a two-parameter model – with mean (μ) and dispersion (α) parameters. How then can the negative binomial be a GLM, which is based on one-parameter models from the exponential family of distributions? Simply enter the dispersion parameter as a constant. Stata, SAS, and R each model the negative binomial as a GLM but estimate the dispersion parameter in a subroutine outside the GLM algorithm, inserting the result back into the IRLS algorithm as if it had been estimated as a GLM. This is a very nice maneuver. All three software applications, however, as well as SPSS, optionally allow users to specify a value for the dispersion if they wish. Stata's **nbreg** automatically estimates the dispersion.

Count data are often found that cannot have zero counts (e.g., hospital length of stay data). However, an analyst is more likely to have data with more zeros than are allowed by Poisson or negative binomial distributional assumptions. Zero-truncated and zero-inflated models have been designed for this purpose. Both negative binomial and Poisson data can be truncated to the left, right, or somewhere in the middle of a distribution of counts. General truncated models exist to adjust the Poisson and negative binomial models for the truncation. Adjustment means that the PDF is amended so that even with an alteration in the original data, the probabilities of the individual component observations in the distribution still sum to one (1). These models, plus the other models that will be examined in the remainder of the book, are all based on some form of the Poisson, some mixture of the Poisson, or an extension of some mixture of Poisson and other GLM distributions.

Finally, I should mention the existence of the *canonical* negative binomial. This model directly derives from the negative binomial PDF. The traditional NB2 negative binomial does not; neither do the other negative binomial models we discuss.

A canonical regression model is nearly always considered a GLM model for which the link and variance functions derive directly from the PDF of the model. For the Poisson PDF, the exponential form of the log-likelihood is equation (1.14):

$$\mathcal{L}\left(\mu; y\right) = \sum_{i=1}^{n} \left\{y_i \log\left(\mu_i\right) \quad - \quad \mu_i \quad - \quad \log(y_i!)\right\}$$

```
           --------        -------     ----------
            link          cumulant    normalization
```

Recall that the exponential family form for count models is $\Sigma \{y\theta - b(\theta) - c(.)\}$, where y is the response count variable being modeled, θ is the link, $b(\theta)$ the cumulant, and $c(.)$ the normalization term, which guarantees that the distribution sums to 1. For the Poisson model, the canonical link is $\log(\mu)$ and the cumulant μ. The first derivative of the cumulant with respect to θ is the mean; the second derivative is the variance. For the Poisson, the result is μ for both the first and second derivatives with respect to θ. This is solved using the chain rule in calculus. Because the Poisson link is $\log(\mu)$, it is the canonical function. For the traditional NB2 negative binomial, the link is also $\log(\mu)$. We want that to be the case so that it can properly model overdispersed Poisson data.

I refer you to equation (5.2), which is the negative binomial log-likelihood function in exponential family form. The canonical link is $\log(\alpha\mu/(1 + \alpha\mu))$, which clearly is not $\log(\mu)$. Given the fact that the NB2 negative binomial is a "noncanonical" model, its standard errors are based on what are called the *expected information matrix* (EIM). Standard errors of canonical models are based on the *observed information matrix* (OIM). Standard errors from the OIM are more accurate than those from the EIM when a model has less than 30 observations. Note that because of this problem, the Stata **glm** command and SAS **Genmod** procedure by default estimate noncanonical models using OIM but allow an option to estimate a model using EIM, or a combination of both. See Hilbe (2011) for a complete discussion of OIM, EIM, and their relationship. Also see the same source for a discussion on the canonical negative binomial. It is to my knowledge the only source on the subject. Our discussion has taken us into the topic of the following section, to which we now turn.

5.2 Negative Binomial Model Assumptions

The traditional negative binomial model, NB2, has the same distributional assumptions as the Poisson distribution, with the exception that it has a

TABLE 5.1. Mean Dispersion Variance Relationships					
M	α	$\mu(1 + \alpha\mu)$	μ	α	$\mu(1 + \alpha\mu)$
.5	.5	0.625	5	.5	17.5
.5	1	0.75	5	1	30
.5	2	1.0	5	2	55
.5	5	1.75	5	5	130
1	.5	1.5	10	.5	60
1	1	2.0	10	1	110
1	2	3.0	10	2	210
1	5	6.0	10	5	510

second parameter – the dispersion parameter – which provides for a wider shape to the distribution of counts than is allowed under Poisson assumptions. The Poisson assumption of equidispersion means that the values of the mean and variance are the same. For the NB2 negative binomial, two parameters affect the variance over that of the mean – the dispersion parameter (α) and square of the mean (μ^2). With greater values of the negative binomial mean come much greater values of the variance. Table 5.1 provides comparative values of the negative binomial mean and variance. Observe how fast the variance increases with increases in the mean. Recall that, for the Poisson distribution, mean = variance ($\mu = \mu$):

> *The negative binomial allows us to model a far wider range of variability than the Poisson.*

The negative binomial distribution can be mathematically derived from the binomial distribution, as well as from the geometric and from a mixture of the Poisson and gamma distributions. For the purpose of statistical modeling, it is the Poisson-gamma mixture interpretation that is predominant.

The negative binomal model is nearly always used to estimate the parameters of overdispersed Poisson data. As mentioned earlier, there are two primary parameterizations of the negative binomial distribution that have been used for this purpose. The first is the standard parameterization, which has a variance function of $\mu + \alpha\mu^2$ or $\mu(1 + \alpha\mu)$. The second parameterizaton, termed NB1 or the linear negative binomial model, has a variance of $\mu + \alpha\mu$ or $\mu(1 + \alpha)$. We discussed the NB1 briefly when examining the quasi-likelihood Poisson model, which has a variance defined as $\phi\mu$. The

NB1 is similar to the quasi-likelihood Poisson in that ϕ can take the term "$1-\alpha$." The difference, however, is that, for the NB1 model, α is estimated as a second parameter – as is the NB2 negative binomial. It is not a constant. If it were, the model would be quasi-likelihood. It is important to remember, though, that statisticians also use the term "quasi-likelihood" when referring to a GLM model that does not specifically derive from an underlying probability function. Poisson or negative binomial models with scaled or robust standard errors, for example, are also referred to as quasi-likelihood models since the standard errors are not directly based on the second derivative of the model log-likelihood, which are called "model" standard errors. The use of the term "quasi-likelihood" in a specific situation must relate to the context in which it is being applied.

Our attention now will focus on the NB2 negative binomial since it is by far the most widespread understanding of negative binomial regression. The negative binomial probability distribution function and associated log-likelihood function are considerably more complex than the Poisson. The probability distribution can be expressed in a variety of ways, with a common parameterization appearing as

$$f(y; \mu, \alpha) = \begin{pmatrix} y_i + \frac{1}{\alpha} - 1 \\ \frac{1}{\alpha} - 1 \end{pmatrix} \left(\frac{1}{1 + \alpha\mu_i} \right)^{\frac{1}{\alpha}} \left(\frac{\alpha\mu_i}{1 + \alpha\mu_i} \right)^{y_i} \tag{5.1}$$

The log-likelihood functions can be given in both μ and exponential mean formats:

$$\mathcal{L}(\mu; y, \alpha) = \sum_{i=1}^{n} y_i \log\left(\frac{\alpha\mu_i}{1 + \alpha\mu_i} \right) - \frac{1}{\alpha} \log\left(1 + \alpha\mu_i\right) + \log\Gamma\left(y_i + \frac{1}{\alpha} \right)$$
$$- \log\Gamma\left(y_i + 1\right) - \log\Gamma\left(\frac{1}{\alpha} \right) \tag{5.2}$$

$$\mathcal{L}(\beta; y, \alpha) = \sum_{i=1}^{n} \left\{ y_i \log\left(\frac{\alpha \exp(x_i'\beta)}{1 + \alpha \exp(x_i'\beta)} \right) - \left(\frac{1}{\alpha} \right) \log\left(1 + \alpha \exp(x_i'\beta)\right) \right.$$
$$\left. + \log\Gamma\left(y_i + \frac{1}{\alpha} \right) - \log\Gamma\left(y_i + 1\right) - \log\Gamma\left(\frac{1}{\alpha} \right) \right\} \tag{5.3}$$

In both forms of log-likelihood, there are three terms with a log-gamma function, all from the normalization term, which is the first term of the PDF. Note that they do not include the mean parameter, μ, or $\exp(x\beta)$, as part of their definition, but do include the dispersion parameter, α, which is to be estimated as well as the mean.

When the negative binomial is estimated using a full maximum likelihood algorithm, both μ and the dispersion parameter α are estimated. When estimated using a generalized linear model algorithm, only μ is estimated; α must be inserted into the algorithm as a constant. Stata's **glm**, R's **glm**, and SAS's **Genmod** procedures all allow the user to provide a value for the negative binomial dispersion parameter. The user can experiment, finding a value that results in the negative binomial dispersion statistic being as close to 1.0 as possible. This value is the correct value of the dispersion parameter, but the method is tedious. **glm** and **Genmod** provide an option that calculates maximum likelihood estimates of α in a subroutine, returning the value to the main estimating algorithm as a constant. The result is that both GLM procedures can produce full maximum likelihood estimates of the coefficients and of α. The developers of R's **glm** function did not provide such capability, but a function called **nb.glm** was authored that calculates the dispersion parameter in the same manner as **glm** and **Genmod**. The caveat, though, is that the developers of **nb.glm** parameterize the dispersion parameter in a manner different from that used in Stata, SAS, SPSS, Limdep, and other major software. It estimates θ (*theta*, not the location parameter or link function meaning of θ discussed earlier), defined as

$$\theta = 1/\alpha$$

Every instance of α in equations (4.1)–(4.3) is inverted.

When the value of α approaches 0, the model is Poisson. For θ, when θ approaches infinity, the model is Poisson. The meaning of θ entails that it has lower values as there is greater variation in the data. As θ approaches 0, the more correlation there is in the data. As θ approaches infinity, the data are more Poisson. Users of R must be careful in how they understand the **glm.nb** dispersion parameter since it is the inverse of α. *I will use the standard direct parameterization of the negative binomial dispersion parmeter (α) when discussing count models unless specifically indicated otherwise.* Note that R functions exist that estimate α rather than θ; for example, the **nbinomial** function available in the **msme** and **COUNT** packages. We discuss this later.

When a Poisson model is overdispersed, the Poisson dispersion statistic, Pearson Chi2/$(n-r)$, is greater than 1, and the negative binomial value of α is greater than 0. A true Poisson model has a Poisson dispersion statistic of 1 and negative binomial dispersion parameter of 0. Repeating from earlier, division by zero prohibits α from having a zero value in the negative binomial PDF or

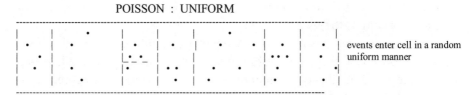

POISSON : UNIFORM

events enter cell in a random uniform manner

FIGURE 5.1. How events can enter or exist within panels of areas or time periods for the Poisson distribution.

in the log-likelihood. However, α can approach 0 (e.g., 0.00001), depending on the software. There is not a one-to-one linear relationship between the Poisson dispersion statistic and negative binomial dispersion parameter. The relationship varies depending on the number and type of predictors, as well as extradispersion. But the relationship is still a positive one.

Finally, recall that in Section 2.6 I explained that, for the Poisson distribution, events enter or exist independently and uniformly within given areas, spaces, or time periods. That is, the events being counted in areas and time periods are described by the uniform distribution. Although this situation does occur in the real world at times, the usual situation is that events occur in nonuniform ways. The negative binomial distribution, as a Poisson-gamma mixture, assumes that events entering into areas or periods are best described according to a gamma distribution. Since uniformly distributed counts of events are equidispersed – where the mean and variance are the same – events entering such areas or periods in nonuniform ways are extradispersed. More often, they are overdispersed. The negative binomial model is a general statistical method that can be used to adjust a count model for overdispersion.

Figures 5.1 and 5.2 give a pictorial view as to how events can enter or exist within panels of areas or time periods. For the Poisson distribution, events are independent and are uniformly entered into each panel or cell. For

NEGATIVE BINOMIAL: POISSON-GAMMA

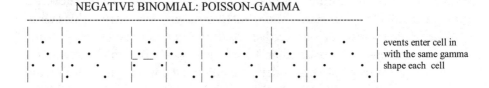

events enter cell in with the same gamma shape each cell

FIGURE 5.2. Same information as in Figure 5.1, but for the negative binomial distribution.

the negative binomial distribution, events occur with a gamma distribution, which is very pliable in the range of shapes it can take. The caveat, however, is that each panel has the same shape (should be drawn to have the same shape). In real life, events are associated with panels in different ways. To model this situation, a simple negative binomial model will not be sufficient. A third parameter may be necessary to describe how a shape – even a gamma shape – changes across panels. To adjust for such a situation, an adjustment to the model standard errors must be made. Most statisticians recommend that robust or empirical standard errors be employed in the presence of heterogeneity or variation in excess of what can be accommodated using standard negative binomial techniques. In fact, I can give this advice:

> *Unless your Poisson or negative binomial model is well fitted and meets its respective distributional assumptions, use robust or empirical standard errors as a default.*

To summarize, the assumptions we make for using a negative binomial (NB2) model are as follows:

- The response, y, is a count consisting of nonnegative integers.
- As the value of μ increases, the probability of 0 counts decreases.
- y must allow for the possibility of 0 counts.
- The fitted or predicted variable, μ, is the expected mean of the distribution of y.
- The variance is closely approximated as $\mu + \alpha\mu^2$ or $\mu(1 + \alpha\mu)$.
- A foremost goal of NB regression is to model data in which the value of the variance exceeds the mean, or the observed variance exceeds the expected variance.
- A well-fitted NB model has a dispersion statistic approximating 1.0 and an AIC/BIC and log-likelihood statistic less than alternative count models.
- The model is not misspecified.
- The number of predicted counts is approximately the same as the number of observed counts across the distribution of y.

5.2.1 A Word Regarding Parameterization of the Negative Binomial

A caveat should be given regarding the parameterization of the negative binomial distribution and model. R, and recently several books that use R

TABLE 5.2. R: Comparison of Direct versus Indirect Dispersion			
medpar	los ~ white+ hmo+type2+ type3	los ~ white+ hmo	value with + dispersion
Poi dispersion	6.260405	7.739524	>>>
DIRECT NB dispersion	1.091024	1.328627	>>>
NB *alpha*	0.445757	0.484603	>>>
INDIRECT NB dispersion	0.254205	0.359399	>>>
NB *theta*	2.243376	2.063547	<<<

when discussing and using negative binomial regression, employ an indirect parameterization of the mean and dispersion parameter.

Stata, SAS, SPSS, Genstat, Limdep, and most major commercial statistical software parameterize the negative binomial with a direct relationship between the mean and dispersion. That is, the greater variability or correlation there is in the data, the greater the value the dispersion parameter will have as well. The variance function for this parameterization of negative binomial is $\mu + \alpha\mu^2$.

Recall that the mean and variance of the Poisson model are both μ, which is the underlying basis for the meaning of equidispersion. A Poisson model is overdispersed if there is more variability in the data than the mean, or the events we are counting do not occur at a uniformly random rate in each time period or area in our data. The negative binomial model has a variance function that allows for modeling overdispersed Poisson data. Table 5.2 provides values demonstrating the difference in direct and indirect parameterizations.

Remember, there is a clear distinction between the dispersion statistic and dispersion parameter. The dispersion statistic is an indicator of excess variability, or lack of it, in the model being fit. The dispersion parameter is a measure of the adjustment made to the overdispersion in the model without the parameter. For example, the negative binomial dispersion parameter shows the adjustment made to the Poisson model. A generalized negative binomial, which we discuss later in the book, adjusts for variability in excess of the distributional assumptions of the negative binomial model. For instance, recall that Poisson dispersion reflects events entering into areas or periods in a uniform manner. The negative binomial then gave a gamma shape to how events enter into periods or areas in y. But the limitation of the negative binomial is its assumption that events enter every period or area in the data in the same way. Events enter with the same type of gamma shape. A three-parameter negative binomial allows for different gamma shapes for each period or area in y. It adjusts or describes the distribution of y in more detail.

Table 5.2 shows that the Poisson dispersion statistic rises when there is increased variability added to the data. Here we dropped two statistically significant predictors, which will increase the model variability. When a negative binomial model is applied to both the full and reduced models, both the negative binomial dispersion statistic and parameter (*alpha*) values increase as well. For the inverted parameterization of the NB dispersion parameter, the NB dispersion statistic increases but the dispersion parameter, *theta*, decreases. It is the inverse value of the direct relationship. With increased variability, the negative binomial value for *theta* decreases. This seems rather counterintuitive.

Why would an analyst prefer to use an indirect relationship between the mean and dispersion parameter for the negative binomial? Essentially the reason comes from the relationship of the variance and gamma variance. The Poisson variance is μ and the gamma variance μ^2/v. Adding the two together produces $\mu + \mu^2/v$. Of course, one can argue that the second term can also appear as $\frac{1}{v}\mu^2$, so that $\alpha = 1/v$, with α being the dispersion parameter. In effect, though, there is no mathematical reason for not inverting v so that the dispersion parameter coheres with the direction of variability in the model. I have to admit that I have had to correct a misinterpretation of the dispersion parameter in numerous manuscripts that I have refereed where R was used by the authors for modeling negative binomial data. Misinterpretations also commonly occur when an analyst who uses Stata, SAS, or another of the major commercial packages evaluates or comments on negative binomial results based on R. However, there are R users such as myself who use **glm.nb** but invert *theta* to *alpha* for interpreting results. Care must be taken when doing this, though, since the Pearson residuals differ as well and they cannot be simply inverted to correspond to a direct relationship.

As a result, I will use the **nbinomial** function when using a negative binomial model for examples in the text. It is a function both in the **msme** and **COUNT** packages on CRAN. Note that **nbinomial** has an option that allows estimation using the indirect relationship as well. The output appears almost identical to that of **glm.nb**, but it also provides a summary of Pearson residuals, the Pearson Chi2 statistic, and dispersion statistic in the output. A likelihood ratio test with a boundary test option is also provided.

5.3 Two Modeling Examples

I will continue using the German health data, **rwm1984**, and the U.S. Medicare data, **medpar**, as example data for estimating negative binomial model parameters. We have found that both sets of data are overdispersed. The data were selected because **rwm1984** has far more zero counts than allowed given both Poisson or negative binomial distributional assumptions and because **medpar** is not capable of having any zero counts. Both of these situations result in overdispersion. However, always keep in mind that overdispersion may come from more than one source. Data can simultaneously have excessive zeros, be structured in clusters, and have missing interaction terms. Adjusting for one source of overdispersion does not thereby absolve other sources from needing adjustment.

There are a host of different ways in which count data can come to an analyst. Analysts with some experience modeling count data will generally recognize those that are overdispersed. A model with no zero counts and a mean value for *y* of 2 to 3 will likely be overdispersed. A model with excessive zero counts will nearly always be overdispersed. If the data are truncated, censored, clustered, or in most any shape other than Poisson, the data will be overdispersed – or sometimes underdispersed.

Underdispersion is typically found when the counts are heavily clumped together – the data are not as dispersed or as variable as would be expected given a Poisson distribution with a specified mean. The negative binomial model cannot adjust for underdispersion. In fact, if an analyst attempts to model data that are underdispersed using a Poisson regression, the software may never converge or may crash. If there is a little excess variability in the model, a negative binomial model might be able to converge with statistical output nearly identical to Poisson, but that's the limit.

5.3.1 Example: rwm1984

For the first example, I'll extend the predictor list for the German health data to include predictors for gender, marital status, and educational level. Gender is indicated with the variable *female*, with female having a value of 1 and male 0. If a patient is *married*, married $= 1$; not married is 0. There are four levels of education, *edlevel1* through *edlevel4*. *edlevel* is a categorical variable but is factored into four separate 0, 1 dummy, or indicator variables.

```
    Level of |
   education |       Freq.      Percent         Cum.
-------------+------------------------------------
 Not HS grad |       3,152       81.36        81.36
     HS grad |         203        5.24        86.60
   Coll/Univ |         289        7.46        94.06
 Grad School |         230        5.94       100.00
```

The default reference level for both Stata and R is the first level. For SAS, it is the highest level, level 4. We use the default, which has over 81% of the observations. In such a case, level 1 is the reasonable reference level since it is both intuitively preferred, being at the base of an ordered group of educational levels, and has by far the most patients. As a caveat, the second level is not significant for most of the models we run on the data, indicating that there is no statistically significant difference in the two groups and their propensity to visit their physician. Visits generally indicate sickness, but not always. If we assume that it does, we cannot differentiate instances of new sickness from treatment for continuing problems. It's best to maintain the visit interpretation.

Next we should look at the distribution of the response variable, *docvis*, which indicates how many visits were made by a patient to the doctor in 1984. We did this for visits 0–7 in Section 2.4; you are invited to look back at that section. To summarize, though, there are 1611 subjects in the study with an observed frequency of 0 counts, which is 41.58% of the data set. We calculated the predicted value in Section 2.4 as well, which, given a Poisson distribution, is 4.2%, nearly ten times less than observed. To calculate what is expected, we do the following:

```
. sum docvis

    Variable |        Obs        Mean    Std. Dev.       Min        Max
-------------+--------------------------------------------------------
      docvis |       3874    3.162881    6.275955          0        121
```

With a mean of 3.16, the expected percentage of 0 counts is

```
. di exp(-3.162881)* (3.162881^0)/exp(lnfactorial(0))
.04230369
```

or an expected 164 patients (3874 * .0423) with no visits to a physician throughout the year.

Stata allows the user to create a global variable containing a list of variables. A global variable is used with a preceding $ sign. We first refer to the expanded model of these data that we developed in Section 2.4, observing a dispersion statistic of over 11.0 and AIC and BIC statistics at approximately 31,000:

```
. use rwm1984
. global xvar "outwork age female married edlevel2 edlevel3 edlevel4"
```

Because of the overdispersion, we employ a robust variance estimator. It is likely that it will accompany us throughout most of the models that will later be examined. With nearly 4000 observations in our model, a dispersion statistic of 11.2 is quite high:

```
. glm docvis $xvar, fam(poi) vce(robust) nolog nohead
```

docvis	Coef.	Robust Std. Err.	z	P>|z|	[95% Conf. Interval]	
outwork	.26473	.0934473	2.83	0.005	.0815767	.4478834
age	.0221169	.002909	7.60	0.000	.0164153	.0278185
female	.2615599	.0914541	2.86	0.004	.0823132	.4408065
married	-.128839	.0871441	-1.48	0.139	-.2996382	.0419602
edlevel2	-.0743016	.1043152	-0.71	0.476	-.2787557	.1301524
edlevel3	-.1825212	.0972583	-1.88	0.061	-.373144	.0081016
edlevel4	-.2644094	.1556778	-1.70	0.089	-.5695323	.0407135
_cons	.0156058	.1801753	0.09	0.931	-.3375313	.3687428

Married and all educational levels fail to significantly contribute to an understanding of visits to the doctor. With the large amount of overdispersion, though, employing a negative binomial model, or later a zero-inflated count model, may result in these predictors being significant.

I have found that the user-authored **countfit** command (Long and Freese 2006) is an excellent means of obtaining an overview of the differences between the Poisson and negative binomial models. The incidence rate ratios (IRR) and standard errors are listed for all predictors for both models in a side-by-side manner to assist comparison:

```
. countfit docvis $xvar, prm nbreg max(12) /* PRM=Poisson regression model */
                                           /* NBRM = NB regression model */
```

Variable	PRM	NBRM
docvis		
1=not working; 0=working	1.303	1.314
	12.31	4.59
Ages 25-64	1.022	1.024
	26.08	9.63
1=femaie; 0=male	1.299	1.372
	12.33	5.60
Married=1; Single=0	0.879	0.826
	-5.91	-2.92
edlevel==HS grad	0.928	0.892
	-1.76	-0.96
edlevel==Coll/Univ	0.833	0.823
	-4.59	-1.91
edlevel==Grad School	0.768	0.696
	-5.50	-3.14
Constant	1.016	0.987
	0.35	-0.11
lnalpha		
Constant		2.251
		26.13
Statistics		
alpha		2.251
N	3874	3874
ll	-1.55e+04	-8306.328
bic	31071.122	16687.015
aic	31021.026	16630.657

ll | -1.55e+04 | -8306.328 | < log-likelihood
bic | 31071.122 | 16687.015 | < substantially lower

legend: b/t

Comparison of Mean Observed and Predicted Count

Model	Maximum Difference	At Value	Mean |Diff|
PRM	0.343	0	0.060
NBRM	-0.040	1	0.010

PRM: Predicted and actual probabilities

```
Count   Actual    Predicted    |Diff|   Pearson
-------------------------------------------------
                 <not displayed>
-------------------------------------------------
Sum      0.952      0.999       0.778  7304.642
```

NBRM: Predicted and actual probabilities

```
Count   Actual    Predicted    |Diff|   Pearson
-------------------------------------------------
0        0.416      0.411       0.005    0.251
1        0.116      0.156       0.040   39.735
2        0.114      0.096       0.017   12.287
3        0.091      0.067       0.024   32.969
4        0.055      0.050       0.005    2.077
5        0.043      0.038       0.005    2.673
6        0.036      0.030       0.006    5.213
7        0.015      0.024       0.009   11.730
8        0.022      0.019       0.002    1.225
9        0.012      0.016       0.004    3.528
10       0.012      0.013       0.002    0.720
11       0.009      0.011       0.002    2.143
12       0.011      0.009       0.002    1.494
-------------------------------------------------
Sum      0.952      0.940       0.124  116.046
```

Tests and Fit Statistics

```
PRM            BIC= -936.032 AIC=     8.007  Prefer  Over  Evidence
-------------------------------------------------------------------
  vs NBRM      BIC=-15320.139 dif= 14384.107 NBRM    PRM   Very strong
               AIC=      4.293 dif=     3.715 NBRM    PRM
               LRX2=14392.370 prob=    0.000 NBRM    PRM   p=0.000
-------------------------------------------------------------------
NBRM           BIC=-15320.139 AIC=     4.293  Prefer  Over  Evidence
```

The note on the upper left-hand side of Figure 5.3 informs us that positive deviations show underpredictions. The figure is a graph of Poisson and NB model residuals. Imagine a horizontal line from 0 to 12 at the zero (0) point. Residuals close to the zero line are well fitted. Both Poisson and negative binomial models fit the data well, starting with the sixth visit. Before that, the negative binomial does much better. This same information can be obtained numerically by viewing the actual and predicted probability tables – which are outstanding diagnostic tools.

Although most information we need is provided here, for comparison purposes with later modeling I'll provide a negative binomial model of the data.

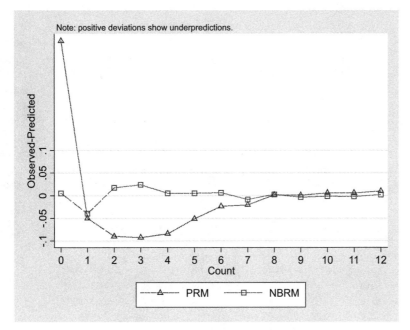

FIGURE 5.3. Graph of Poisson and NB model residuals.

We find that the dispersion is 1.43, indicating that there is more variability in the data than can be adjusted for using a negative binomial model. Recall that we still have the excessive zero counts we have not adjusted for. Note that the dispersion parameter is 2.25, which is fairly high. We should also compare it with the NB1 linear negative binomial model:

```
NB2 - NEGATIVE BINOMIAL (TRADITIONAL)

. glm docvis $xvar, fam(nb ml) vce(robust) nolog

Generalized linear models              No. of obs      =       3874
Optimization     : ML                  Residual df     =       3866
                                       Scale parameter =          1
Deviance         =   3909.433546       (1/df) Deviance =   1.011235
Pearson          =   5517.070211       (1/df) Pearson  =   1.427075
Variance function: V(u) = u+(2.2506)u^2         [Neg. Binomial]
                                       AIC             =   4.292374
Log pseudolikelihood = -8306.328282    BIC             =  -28031.62
```

```
-----------------------------------------------------------------------------
             |              Robust
      docvis |     Coef.   Std. Err.      z    P>|z|     [95% Conf. Interval]
-------------+---------------------------------------------------------------
     outwork |   .2733504   .0812069    3.37   0.001     .1141879    .4325129
         age |   .0232448   .0026847    8.66   0.000     .0179828    .0285068
      female |   .3164156   .0758391    4.17   0.000     .1677737    .4650576
     married |  -.1906226   .0855496   -2.23   0.026    -.3582968   -.0229485
     edlevel2|  -.1139377   .1051687   -1.08   0.279    -.3200647    .0921893
     edlevel3|  -.1948105    .098729   -1.97   0.048    -.3883158   -.0013051
     edlevel4|  -.3628498   .1321642   -2.75   0.006    -.6218868   -.1038127
       _cons |   -.013249   .1520918   -0.09   0.931    -.3113435    .2848456
-----------------------------------------------------------------------------

. qui nbreg docvis $xvar,  vce(robust) nolog
. abich
AIC Statistic   =      4.29289         AIC*n       = 16630.656
BIC Statistic   =     4.298453         BIC(Stata)  = 16687.016
AICH Statistic  =     4.291909         AICH*n      = 16605.557
```

This code captures the log-likelihood from the model, storing it in *llnb2*:

```
. scalar llnb2 = e(ll)
```

It will help in assessing the fit of the model if we can see the relationship of the fit and observed probabilities of *docvis*. Stata code for constructing Figure 5.3 is provided. It is an adaptation of the code in Hilbe (2011). This code can be adapted for any negative binomial model using the **nbreg** command.

Caveat: recall that the **glm** command enters the dispersion parameter into the estimation algorithm as a constant. As such, it does not add the dispersion as an extra parameter when calculating the AIC and BIC degrees of freedom. The full maximum likelihood command, **nbreg**, estimates the dispersion, adding it to the degrees of freedom. There is therefore a slight difference in AIC and BIC values between the two methods. The dispersion parameter should be added to the calculation. I calculated the correct values by quietly running **nbreg** before **abich**.

To run the code in Table 5.3, I suggest downloading the file from the book's web site, typing **doedit** on the Stata command line, and pasting the code into the editor. When ready to run, click "Tools" on the menu and then click on "Execute (do)." The figure will be in the graphics screen.

If the option *disp(constant)* is placed at the end of the second line of code in Table 5.3, an NB1 model will be produced. You can compare the

TABLE 5.3. Stata: Observed versus Predicted Counts for `docvis`

```
qui {
qui nbreg docvis outwork age married female edlevel2 edlevel3 edlevel4
predict mu
local alpha = e(alpha)
gen amu = mu* e(alpha)
local i 0
local newvar "pr`i'"
while `i' <=15 {
local newvar "pr`i'"
qui gen `newvar' = exp(`i'*ln(amu/(1+amu)) - (1/`alpha')*ln(1+ amu)+
lngamma(`i' +1/`alpha') /*
   */  - lngamma(`i'+1) - lngamma(1/`alpha'))
local i = `i' + 1
}
quietly gen cnt = .
quietly gen obpr = .
quietly gen prpr = .
local i 0
while `i' <=15 {
 local obs = `i' + 1
 replace cnt = `i' in `obs'
 tempvar obser
 gen `obser' = (`e(depvar)'==`i')
 sum `obser'
 replace obpr = r(mean) in `obs'
 sum pr`i'
 replace prpr = r(mean) in `obs'
 local i = `i' + 1
}
gen byte count = cnt
label var prpr "NB2 - Predicted"
label var obpr "NB2 - Observed"
label var count "Count"
}
twoway scatter prpr obpr count, c(l l) ms(T d) title("docvis Observed vs
Predicted Probabilities") sub("Negative Binomial") ytitle(Probability of
Physician Visits)
```

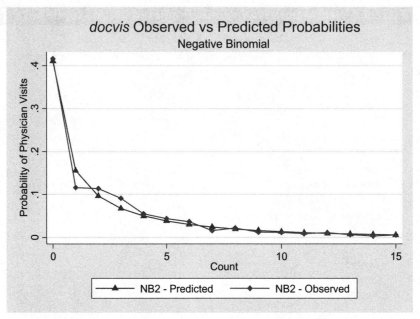

FIGURE 5.4. Negative binomial model: *docvis* observed vs. predicted probabilities.

predicted values for NB1 and NB2 by typing the following on the command line:

```
. rename prpr prnb2
. rename obpr obnb2
. drop mu*
. drop count
<then rerun the code in the dofile editor with the changes you have made
to it>
```

You may adapt the code in Table 5.3 for use with other models. However, you should type "*ereturn list*" to check the *e()* macros that are saved following the execution of the regression to determine if their names are different. If you use the correct macro names, you can adapt the code in Table 5.3 to compare observed versus predicted probabilities across a number of different regression models. R users can usually find the names of saved statistics in the help files.

We have examined the relationship of Poisson and NB2 models, and the observed versus predicted values of the response variable, *docvis* (see Figure 5.4). A remaining comparison can be made between NB2 and NB1:

```
NB1 - LINEAR NEGATIVE BINOMIAL

. nbreg docvis $xvar, nolog vce(robust) disp(constant)

Negative binomial regression                    Number of obs    =        3874
Dispersion            = constant                Wald chi2(7)     =      312.30
Log pseudolikelihood = -8283.0432               Prob > chi2      =      0.0000
-----------------------------------------------------------------------------
             |              Robust
      docvis |     Coef.    Std. Err.      z     P>|z|     [95% Conf. Interval]
-------------+---------------------------------------------------------------
     outwork |   .1861926   .0489325    3.81    0.000     .0902868    .2820985
         age |     .01871   .0018376   10.18    0.000     .0151084    .0223116
      female |   .3081968   .0481739    6.40    0.000     .2137777    .4026158
     married |  -.0430265   .0487428   -0.88    0.377    -.1385605    .0525076
     edlevel2 |   .0533671   .0844128    0.63    0.527     -.112079    .2188132
     edlevel3 |  -.0041829   .0772246   -0.05    0.957    -.1555403    .1471746
     edlevel4 |  -.2383558   .0989037   -2.41    0.016    -.4322035   -.0445081
        _cons |   .1013333   .1021209    0.99    0.321    -.0988199    .3014865
-------------+---------------------------------------------------------------
    /lndelta |   1.976527   .0561385                       1.866498    2.086556
-------------+---------------------------------------------------------------
       delta |   7.217632    .405187                       6.465611    8.057122
-----------------------------------------------------------------------------
. abich
AIC Statistic   =    4.280869        AIC*n       = 16584.086
BIC Statistic   =    4.286432        BIC(Stata)  = 16640.445
AICH Statistic  =    4.279888        AICH*n      = 16558.986

. scalar llnb1 = e(ll)
```

Note that the AIC and BIC statistics are better for the NB1 model. Aside from the AIC and BIC comparative-fit tests, the optimal tool for deciding between NB1 and NB2 is the generalized NB-P negative binomial, discussed in the final section of this chapter. I leave the model diagnostics to you.

R code in Tables 5.4 and 5.5 provides a means to replicate some of the output obtained from using **countfit**. Additional code will be made available on the book's web site as it is produced.

The interpretation of the Poisson and negative binomial models is the same. Any model with a log link function will have the same interpretation. When the coefficients are exponentiated, both regressions will be able to be interpreted as incidence rate ratios, as discussed in Chapter 2, on Poisson regression. You are referred there for the following model:

```
. nbreg docvis $xvar, nolog  vce(robust) irr

Negative binomial regression              Number of obs   =      3874
Dispersion             = mean             Wald chi2(7)    =    231.08
Log pseudolikelihood = -8306.3283         Prob > chi2     =    0.0000

-------------------------------------------------------------------------
             |               Robust
      docvis |      IRR   Std. Err.      z    P>|z|     [95% Conf. Interval]
-------------+-----------------------------------------------------------
     outwork |  1.314361   .106737     3.37   0.001     1.12096    1.54113
         age |  1.023517   .0027479    8.66   0.000    1.018146   1.028917
      female |  1.372201   .1039725    4.18   0.000    1.182828   1.591892
     married |   .8264441  .0707537   -2.23   0.026     .6987797   .9774323
    edlevel2 |   .8923135  .0938496   -1.08   0.279     .7260923  1.096587
    edlevel3 |   .8229906  .0812487   -1.97   0.048     .6782051   .9986852
    edlevel4 |   .6956908  .0919757   -2.74   0.006     .5368845   .9014707
       _cons |   .9868381  .150062    -0.09   0.931     .7325028  1.329482
-------------+-----------------------------------------------------------
    /lnalpha |   .8112091  .037307                      .7380888   .8843295
-------------+-----------------------------------------------------------
       alpha |  2.250628   .0839641                     2.091934    2.42136
-------------------------------------------------------------------------
```

TABLE 5.4. R: `rwm1984` Modeling Example

```
# make certain the appropriate packages are loaded
library(COUNT); data(rwm5yr); rwm1984 <- subset(rwm5yr, year==1984)
# USING glm.nb
summary(nbx <- glm.nb(docvis ~ outwork + age + married + female +
      edlevel2 + edlevel3 + edlevel4, data=rwm1984))
exp(coef(nbx)); exp(coef(nbx))*sqrt(diag(vcov(nbx)))
exp(confint.default(nbx))
alpha <- 1/nbx$theta; alpha; P__disp(nbx)
modelfit(nbx)
xbnb <- predict(poi1); munb <- exp(xbnb)
# expected variance of NB model (using alpha where alpha=1/theta)
mean(munb)+ (1/nbx$theta)*mean(munb)^2
round(sqrt(rbind(diag(vcov(nbx)), diag(sandwich(nbx)))), digits=4)
# USING nbinomial
summary(nb1 <- nbinomial(docvis ~ outwork + age + married + female +
      edlevel2 + edlevel3 + edlevel4, data=rwm1984))
modelfit(nb1)
```

TABLE 5.5. R: `rwm1984` Poisson and NB2 Models

```
library(COUNT);library(msme)
data(rwm5yr);rwm1984 <- subset(rwm5yr, year==1984)
# POISSON
poi <- glm(docvis ~ outwork + age + married + female +
                   edlevel2 + edlevel3 + edlevel4,
                   family = poisson, data = rwm1984)
summary(poi)
#NB2
summary(nb1 <- nbinomial(docvis ~ outwork + age + married + female +
         edlevel2 + edlevel3 + edlevel4, data=rwm1984))
# NB1
library (gamlss)
summary(gamlss(formula = docvis ~ outwork + age + married + female +
     edlevel2 + edlevel3 + edlevel4, family = NBII, data = rwm1984))
```

I will work through the interpretations of the preceding table, with the understanding that there are alternative ways of expressing essentially the same thing. If you are interpreting a model based on coefficient values, it is wise to exponentiate them before attempting an interpretation. Exponentiated coefficients are called incidence rate ratios. When the response is binary, an

TABLE 5.6. Interpretation of German Health Care Parameter Estimates

Patients who are out of work visit the doctor some 31% more often than working patients.

For each one-year increase in age, a patient is expected to go to the doctor a little over 2% more often.

Females see the doctor some 37% more often than males.

Single patients are expected to go to the doctor some 21% more often than married patients. Note that $1/.8264441 = 1.21000$. Or we can affirm that married patients are expected to see the doctor some 17.5% less often than single patients. I prefer to interpret the relationship in a positive manner (i.e., "more than").

Patients who have a college or university education are expected to go to the doctor some 17.5% less often than those who have not been to college. Because *edlevel2* is not significant, *edlevel1* and *edlevel2* have been merged. Edlevel1–2 is the reference level.

Patients with at least some graduate school education were about 30% less likely to see a doctor (in 1984) than patients who never went to college.

exponentiated coefficient is a risk ratio, but counts may be interpreted as risks as well, if it makes sense.

For each of the interpretations in Table 5.6, the predictors not being interpreted are held to a constant value, typically at their mean value. The important thing is that the interpretation be adjusted by the other predictors in the model. The adjustment results in the interpretation they are given. (Note: This is because coefficients are partial derivatives.) I will not express the common phrase "holding the values of the other predictors at a constant." It is understood.

5.3.2 Example: `medpar`

The **medpar** data, as you may recall, are part of the U.S. Medicare data for Arizona in 1991. A specific diagnostic group is represented. The response, length of stay (*los*), represents how long a patient was registered in a hospital for a specific disease. *type* is a categorical variable that has been factored into three levels or categories and is used to explain *los*. The categories are defined in the following table.

```
. tab type
       Type of |
      admission |      Freq.      Percent        Cum.
----------------+-----------------------------------
  Elective Admit |      1,134       75.85        75.85
    Urgent Admit |        265       17.73        93.58
 Emergency Admit |         96        6.42       100.00
----------------+-----------------------------------
          Total |      1,495      100.00
```

White indicates that the patient identified themselves as Caucasian. *hmo* is an acronym for health maintenance organization. If a patient belongs to one, *hmo* = 1. If not, they are coded 0.

A Poisson model of the data, with robust standard errors, can be given as follows:

```
. glm los hmo white type2 type3, nolog fam(poi) vce(robust)

Generalized linear models              No. of obs       =      1495
Optimization     : ML                  Residual df      =      1490
                                       Scale parameter  =         1
```

```
Deviance        =   8142.666001            (1/df) Deviance =   5.464877
Pearson         =   9327.983215            (1/df) Pearson  =   6.260391
                                           AIC             =   9.276131
Log pseudolikelihood = -6928.907786        BIC             =  -2749.057
------------------------------------------------------------------------
             |               Robust
     los |     Coef.     Std. Err.      z    P>|z|     [95% Conf. Interval]
--------+---------------------------------------------------------------
     hmo |  -.0715493    .0517323    -1.38   0.167    -.1729427    .0298441
   white |  -.153871     .0833013    -1.85   0.065    -.3171386    .0093965
   type2 |   .2216518    .0528824     4.19   0.000     .1180042    .3252993
   type3 |   .7094767    .1158289     6.13   0.000     .4824562    .9364972
   _cons |   2.332933    .0787856    29.61   0.000     2.178516    2.48735
------------------------------------------------------------------------
. abich
AIC Statistic   =   9.276131         AIC*n      = 13867.815
BIC Statistic   =   9.280208         BIC(Stata) = 13894.365
AICH Statistic  =   9.274205         AICH*n     = 13854.255
```

Note the high dispersion statistic at 6.26. Also, robust or empirical standard errors were given the model because of the overdispersion. The model without robust SEs had all *p*-values as statistically significant. It is noteworthy that the data do not have 0 counts. We will come back to this when we evaluate zero-truncated models.

We next tabulate *los*. Note the comparatively greater number of 1 counts. Some 8.5% of patients stayed only a single night in the hospital:

```
. tab los

  hospital |
 Length of |
      Stay |       Freq.     Percent      Cum.
-----------+-----------------------------------
         1 |        126        8.43        8.43
         2 |         71        4.75       13.18
         3 |         75        5.02       18.19
         4 |        104        6.96       25.15
         5 |        123        8.23       33.38
           .          .           .
        74 |          1        0.07       99.87
        91 |          1        0.07       99.93
       116 |          1        0.07      100.00
```

Run a negative binomial model, comparing it to the earlier Poisson models we made of the data:

```
. glm los hmo white type2 type3, nolog fam(nb ml) vce(robust)

Generalized linear models              No. of obs       =      1495
Optimization     : ML                  Residual df      =      1490
                                       Scale parameter  =         1
Deviance         =   1568.14286        (1/df) Deviance  =  1.052445
Pearson          =   1624.538251       (1/df) Pearson   =  1.090294
Variance function: V(u) = u+(.4458)u^2 [Neg. Binomial]
                                       AIC              =  6.424718
Log pseudolikelihood = -4797.476603    BIC              = -9323.581

-----------------------------------------------------------------------
             |              Robust
       los |    Coef.   Std. Err.      z    P>|z|    [95% Conf. Interval]
-------+---------------------------------------------------------------
       hmo | -.0679552   .0513265   -1.32   0.186   -.1685533    .0326429
     white | -.1290654   .0710282   -1.82   0.069   -.2682782    .0101473
     type2 |   .221249   .0530361    4.17   0.000    .1173001    .3251978
     type3 |  .7061588   .1157433    6.10   0.000    .4793061    .9330116
     _cons |  2.310279   .0689142   33.52   0.000     2.17521    2.445348
-----------------------------------------------------------------------

. abich
AIC Statistic   =    6.426055        AIC*n       = 9606.9531
BIC Statistic   =    6.432411        BIC(Stata)  = 9638.8125
AICH Statistic  =    6.42592         AICH*n      = 9589.0547
```

The negative binomial is a substantial improvement over the Poisson model. The dispersion has been reduced to 1.09, indicating that the data still have more variability than is accounted for using a negative binomial model. Without the robust standard errors, the model SEs tell us that *hmo* is clearly not significant and that *white* likely is not either. Robust standard errors tell us the same story. We may suspect that the model is misspecified, so try checking whether NB2 or NB1 better fits this data using a **linktest**:

```
. linktest

Negative binomial regression          Number of obs    =      1495
                                       LR chi2(2)       =    118.46
Dispersion      = mean                 Prob > chi2      =    0.0000
Log likelihood = -4797.2636            Pseudo R2        =    0.0122
```

```
----------------------------------------------------------------------
     los |      Coef.   Std. Err.       z    P>|z|     [95% Conf. Interval]
---------+------------------------------------------------------------
    _hat |  -.2952803    1.986614    -0.15    0.882    -4.188973    3.598412
  _hatsq |   .2598939    .3982465     0.65    0.514    -.5206548    1.040443
   _cons |    1.59086    2.446458     0.65    0.516     -3.20411     6.38583
---------+------------------------------------------------------------
/lnalpha |  -.8083432    .0444598                      -.8954828   -.7212036
---------+------------------------------------------------------------
   alpha |   .4455957    .0198111                       .4084104    .4861668
----------------------------------------------------------------------
Likelihood-ratio test of alpha=0:  chibar2(01) = 4260.59 Prob>=chibar2 =
0.000
```

The *_hatsq* predictor is not significant, indicating that the model is not misspecified. An NB1 test is attempted next.

To check the NB1 parameterization, we run a model, check the AIC and BIC statistics, and run a **linktest**. To save space, we will not display the regression results but only the summary statistics of interest. Placing a *qui* in front of a Stata command suppresses the display to screen:

```
. qui nbreg los hmo white type2 type3, nolog disp(const)
. abich
AIC Statistic   =    6.471913          AIC*n      = 9675.5107
BIC Statistic   =    6.478269          BIC(Stata) = 9707.3701
AICH Statistic  =    6.471778          AICH*n     = 9657.6123

. linktest
```

```
----------------------------------------------------------------------
     los |      Coef.   Std. Err.       z    P>|z|     [95% Conf. Interval]
--------+-------------------------------------------------------------
    _hat |  -8.003606    4.769589    -1.68    0.093    -17.35183    1.344617
  _hatsq |   1.987063    .9933492     2.00    0.045     .0401342    3.933992
   _cons |   10.15396    5.700143     1.78    0.075    -1.018113    21.32604
  _hatsq |   1.987063    .9933492     2.00    0.045     .0401342    3.933992
----------------------------------------------------------------------
```

The **linktest** fails, indicating that the NB1 model is misspecified. In addition, the AIC and BIC statistics are greater than for NB2, and of course very much greater than with Poisson.

At this point, it is clear that we need to determine whether the lack of 0 counts in the model affects the dispersion. It's greater than it should be, but not bad.

TABLE 5.7. *Medpar* Data – IC Test Results			
	POISSON	NB1	NB2
AIC*n	13867.815	9675.5107	9606.9531
BIC	13894.365	9707.3701	9638.8125
AICH*n	13854.255	9657.6123	9589.0547

Variable		Obs	Mean	Std. Dev.	Min	Max
los		1495	9.854181	8.832906	1	116

The mean of *los* is 9.85, indicating that, on average, patients were in the hospital some 10 days. Because the mean is so high, it is unlikely that lack of 0 counts will make a difference. With a mean of 9, the percentage of 0 counts expected is near 0. Also look at the standard deviation, 8.83. The variance is the square of the standard deviaton – 78. This value clearly violates the Poisson assumption of equidispersion. We will explore other alternatives for these data in subsequent chapters as we look at more complex modeling strategies.

5.4 ADDITIONAL TESTS

5.4.1 General Negative Binomial Fit Tests

There are several tests we have previously discussed that relate to testing the fit of negative binomial models. In general, if a negative binomial model has a substantially lower AIC and BIC set of statistics than other models, then it can be regarded as the better-fitted model. By set of AIC and BIC statistics I mean the values displayed in the **abic** or **abich** command output when using Stata (author commands), or **modelfit** when using R. **modelfit** is a function in the **COUNT** package, where it is associated with the **irls** and **ml** group of functions as well as with R's **glm** and **glm.nb** functions. It does not currently work with **nbinomial**.

Negative binomial regression is nearly always used to model overdispersed Poisson data. What happens, though, if the value of *alpha* (α) is close to 0? The Poisson model can be thought of as a negative binomial with 0 dispersion (i.e., with $\alpha \approx 0$, where values slightly greater than 0 can be within the confidence interval for the Poisson dispersion). In Section 4.2.2, we demonstrated the use of a one-sided boundary likelihood ratio test to be used following estimation of a negative binomial model, testing whether the dispersion parameter, α, is statistically different from 0. You are referred to that discussion. We also presented several tests for overdispersion in Section 4.3. You are also referred there.

In the following two subsections, we test negative binomial models by employing different types of regression models. Both of the extended negative binomial models we consider here have been referred to as generalized negative binomial models. The first, NB-P, is a generalized negative binomial model since it adds an additional parameter to the base NB2/NB1 model. The second, NB-H, is not a generalized negative binomial in the usual sense, although it is called that by Stata.

Both models test the base negative binomial for different purposes. The NB-P model is primarily used to decide between using an NB2 model or an NB1 model. It does this by parameterizing the negative binomial exponent. The heterogeneous negative binomial, NB-H, parameterizes the dispersion parameter, with the goal of determining which predictors add or adjust for excess variability in the data. This in turn produces overdispersion in the data.

5.4.2 Adding a Parameter – NB-P Negative Binomial

Earlier, we introduced the NB1 or linear negative binomial model. There are times when it fits the data more optimally than NB2. William Greene (2008) of New York University (and author of the Limdep econometrics software program) developed a variation of the negative binomial model that allows the exponent of the second term of the variance to be parameterized.[1] The effect of doing so is that the dispersion statistic is allowed to vary across observations:

Poisson – equidispersion; variance does not vary from mean

NB – dispersion parameter allows a constant dispersion value for all observations

NB-P – a second dispersion parameter allows the dispersion to vary across observations

---

[1] Although Greene was the first to actually develop software for parameterizing the exponent of the negative binomial variance, this parameterization is mentioned in Cameron and Trivedi (1998), Winkelmann (2008), and described in Hilbe and Greene (2008).

To explain, recall that the variance functions of the NB1 and NB2 models are, respectively,

$$\text{NB1} \quad \mu + \alpha\mu \text{ or } \mu + \alpha\mu^1 \tag{5.4}$$

and

$$\text{NB2} \quad \mu + \alpha\mu^2 \tag{5.5}$$

The difference rests in the value of the exponent. To choose between them, Greene let the exponent become a parameter to be estimated, called ρ, for power. ρ is the Greek letter rho. The model is called, appropriately, NB-P:

$$\text{NB-P} \quad \mu + \alpha\,\mu^\rho \tag{5.6}$$

The derivation of the NB-P probability and log-likelihood functions can be found in Hilbe (2011) and in the Limdep Reference Manual. As an example, we use the German health data that have been used throughout the text. First, let's begin by creating a global variable, called *xvar*, that contains our predictor list:

```
. use rwm1984
. global xvar "outwork age female married edlevel2 edlevel3 edlevel4"
. nbregp docvis $xvar, nolog vce(robust)

Negative binomial-P regression              Number of obs   =       3874
                                            Wald chi2(7)    =     186.34
Log pseudolikelihood = -8277.964            Prob > chi2     =     0.0000

-------------------------------------------------------------------------
             |               Robust
     docvis  |     Coef.   Std. Err.      z    P>|z|    [95% Conf. Interval]
-------------+-----------------------------------------------------------
    outwork  |  .2329032   .0632728    3.68   0.000    .1088908    .3569157
        age  |  .0221351   .0024555    9.01   0.000    .0173225    .0269477
     female  |  .3420894   .0594544    5.75   0.000     .225561    .4586178
    married  | -.0799971    .064053   -1.25   0.212   -.2055387    .0455445
   edlevel2  |  .0163448   .0993362    0.16   0.869   -.1783506    .2110403
   edlevel3  | -.0555854   .0937499   -0.59   0.553   -.2393319    .1281611
   edlevel4  | -.2940122   .1187276   -2.48   0.013    -.526714   -.0613103
      _cons  | -.0660212   .1287343   -0.51   0.608   -.3183357    .1862934
-------------+-----------------------------------------------------------
         /P  |  1.363081   .1248427   10.92   0.000    1.118394    1.607769
    /lntheta |  1.545855   .1509675                    1.249964    1.841746
-------------+-----------------------------------------------------------
      theta  |   4.69198   .7083367                    3.490217    6.307539
-------------------------------------------------------------------------
Likelihood-ratio test of P=1:     chi2 =    10.16 Prob > chi2    =    0.0014
Likelihood-ratio test of P=2:     chi2 =    56.73 Prob > chi2    =    0.0000
```

```
. abich
AIC Statistic  =   4.278763          AIC*n     = 16575.928
BIC Statistic  =   4.285488          BIC(Stata) = 16638.549

. scalar llnbp = e(ll)
```

The AIC and BIC values are only a little less than for the NB1 model but are significantly less than in NB2. However, this finding needs to be checked using a likelihood ratio test. The likelihood ratio (LR) test between NB-P and NB2 can be given as follows:

```
LIKELIHOOD RATIO TEST
. di -2*(llnb2 - llnbp)
56.728519

P-VALUE FOR LR TEST
. di chi2tail(1, 56.72852)
5.003e-14
```

The likelihood ratio test of the hypothesis that the model having a ρ of 1.36 is statistically the same as NB2, with a $\rho = 2$, is rejected. Recall that with 1 degree of freedom, any Chi2 test statistic having a value greater than 3.84 will reject the hypothesis that the models being compared are statistically the same. The same holds when comparing NBP and NB1, with $\rho = 1$. These tests inform us that the model is neither NB1 nor NB2.

Recall also that Greene provides us with another test statistic – one to determine whether NB2 is preferable to NB1. It is based on the powers. The resultant statistic is distributed as a *reverse cumulative upper tail Student's* t *test* with 1 degree of freedom:

$$\frac{nbp - nb2}{nbp_{se}} \tag{5.7}$$

Here we have

```
. di (1.363081 -2)/.1248427
-5.1017721

. di ttail(1, -5.1017721)
.93838908
```

Values of p greater than 0.05 are not significantly different. For our models, the data can be modeled using either NB1 or NB2. However, given the lower value of the AIC and BIC statistics, NB1 should likely be preferred.

TABLE 5.8. Major Points: NB-P

- The NB-P model allows the dispersion to vary across observations instead of assuming that it is a constant shape for all observations in the model.
- The estimated power gives an indication of preference to modeling as NB2 or NB1 but allows the model to stand in its own right.
- Likelihood ratio tests are given comparing the NB-P with NB2 and NB1. The test with the lowest Chi2 statistic is generally preferred.

There are some, including myself, who will take the NB-P model results as valuable in themselves, and not merely as a test between NB2 or NB1. If an NB-P test result comes with a substantially lower information criterion value, then I suggest using NB-P as the model of choice. In this case, no such determination can be made between NB1 and NB2, so NB-P should be used as the preferred model.

As of this writing, no NB-P function exists in R. As soon as a function is developed, I will provide information on this book's web site regarding how it may be downloaded.

5.4.3 Modeling the Dispersion – Heterogeneous Negative Binomial

In order to determine the source of over- or underdispersion, it may be helpful to discover how the predictors themselves contribute to the dispersion parameter. There is a statistical procedure designed for exactly this purpose. In particular, I have found it to be invaluable for discovering sources of overdispersion. At the very minimum, knowing that a specific predictor significantly contributes to the dispersion parameter allows the analyst to explore the predictor in more detail and to have additional information about the data for discussion in a research report.

As an example, I will use the Squirrel data set (**nuts**) from Zuur, Hilbe, and Ieno (2013). As originally reported by Flaherty et al. (2012), researchers recorded information about squirrel behavior and forest attributes across various plots in Scotland's Abernethy Forest. The study focused on the following variables:

response *cones* = number of cones stripped by red squirrels per plot

predictor *sntrees* = standardized number of trees per plot

sheight = standardized mean tree height per plot

scover = standardized percentage of canopy cover per plot

The stripped cone count was only taken when the mean diameter of trees was under 0.6 m (*dbh*). I will provide R output as the primary example for this section. I begin by modeling the data with a Poisson model in order to obtain a base model and to determine the Poisson dispersion statistic. R's **glm** *quasipoisson* model provides the dispersion statistic in its output. Poisson does not:

```
R CODE
POISSON
> library(COUNT) : data(nuts)
> nut <- subset(nuts, dbh<.6)
> sntrees <- scale(nut$ntrees)
> sheight <- scale(nut$height)
> scover  <- scale(nut$cover)
> summary(PO <- glm(cones ~ sntrees + sheight + scover,
family=quasipoisson, data=nut))

Coefficients:
            Estimate Std. Error t value Pr(>|t|)
(Intercept)   2.6352     0.1546  17.041   <2e-16 ***
sntrees       0.2739     0.1068   2.566   0.0135 *
sheight       0.2004     0.1734   1.156   0.2536
scover        0.5280     0.2439   2.164   0.0355 *
---
(Dispersion parameter for quasipoisson family taken to be 13.68122)
```

The dispersion statistic is 13.68, indicating substantial Poisson overdispersion. Only the standardized tree height predictor is not significant, but because of the overdispersion we have no idea of the significance status of any of the predictors. A table of the response, *SqCones*, shows that the mean number of cones stripped is 16.49. Five 0 counts is excessive, but the data are fairly uniformly distributed after 5. The excess zero counts may be a reason for overdispersion, but we really have no idea whether that is the case:

```
> table(nut$cones)

 0  1  2  3  4  5  6  7  9 11 12 14 15 16 17 18 20 21 22 25 27 32 35 42 44 47
 5  1  3  4  5  3  1  1  2  2  2  1  1  2  1  2  1  1  1  1  1  1  1  1  1  1
48 55 57 60 61
 1  1  1  1  1
```

```
> summary(nut$cones )
   Min. 1st Qu.  Median    Mean 3rd Qu.    Max.
   0.00    3.50   11.00   16.49   21.50   61.00
```

A negative binomial model is a reasonable one to use at this point in order to determine how much of the variability in the model can be appropriately adjusted:

```
NEGATIVE BINOMIAL
> library(msme)
> summary(NB <- nbinomial(cones ~ sntrees + sheight + scover, data=nut)
       .    .    .
Coefficients (all in linear predictor):
               Estimate    SE     Z        p       LCL    UCL
(Intercept)       2.627 0.142 18.53 1.24e-76  2.34948 2.905
sntrees           0.306 0.161  1.90   0.0573 -0.00945 0.621
sheight           0.158 0.159  0.99    0.322 -0.15463 0.470
scover            0.536 0.179  3.00  0.00274  0.18516 0.886
(Intercept)_s     0.927 0.199  4.65 3.37e-06  0.53584 1.318

Null deviance: 77.26177  on  49 d.f.
Residual deviance: 59.07252  on  46 d.f.
Null Pearson: 58.54636  on  49 d.f.
Residual Pearson: 41.19523  on  46 d.f.
Dispersion: 0.8955485
AIC:  383.8132
```

The Poisson overdispersion in the data is well adjusted by the negative binomial, but the data are actually NB2-underdispersed, with a dispersion of .90. We can determine the source of the Poisson overdispersion by modeling the data with a heterogeneous negative binomial. Rather than the dispersion being the same in each cell of counts, as is the case with standard negative binomial regression, the dispersion here is itself parameterized, and differs from cell to cell. We then might determine which predictors have the most influence on extradispersion:

```
HETEROGENEOUS NEGATIVE BINOMIAL

> summary(HNB <- nbinomial(cones ~ sntrees + sheight + scover,
+     formula2 =~ sntrees + sheight + scover, data=nut, family = "negBinomial",
+     scale.link = "log_s"))
Deviance Residuals:
   Min. 1st Qu.  Median    Mean 3rd Qu.    Max.
-2.7910 -0.9137 -0.3023 -0.3208  0.4419  1.5670
```

```
Pearson Residuals:
     Min.   1st Qu.    Median     Mean  3rd Qu.      Max.
-1.14700  -0.70960  -0.23950  0.03088  0.31550   3.20200

Coefficients (all in linear predictor):
              Estimate     SE       Z        p     LCL    UCL
(Intercept)     2.6147  0.144  18.159  1.1e-73  2.3325  2.897
sntrees         0.2731  0.110   2.478   0.0132  0.0571  0.489
sheight         0.0744  0.144   0.516    0.606 -0.2081  0.357
scover          0.5217  0.158   3.300 0.000967  0.2118  0.832
(Intercept)_s  -0.1950  0.238  -0.821    0.412 -0.6608  0.271
sntrees_s      -0.3834  0.323  -1.186    0.236 -1.0168  0.250
sheight_s       0.3312  0.321   1.033    0.302 -0.2973  0.960
scover_s        0.2723  0.417   0.652    0.514 -0.5459  1.091

Null deviance: 85.07187  on  49 d.f.
Residual deviance: 59.49816  on  43 d.f.
Null Pearson: 45.30266  on  49 d.f.
Residual Pearson: 46.82167  on  43 d.f.
Dispersion: 1.088876
AIC:  385.7116

> exp(coef(HNB))
 (Intercept)        sntrees        sheight         scover (Intercept)_s
  13.6637970      1.3140450      1.0772100      1.6849189      0.8228193
    sntrees_s      sheight_s       scover_s
   0.6815513      1.3925859      1.3130309
```

Exponentiation of the coefficients provides us with incidence rate ratios. Remember, though, that the dispersion parameter is now partitioned across the predictors we have selected to parameterize the dispersion. Here none of the dispersion component predictors appear significant. I suggest that the standard errors be adjusted using a robust variance estimator. I adjusted the H-NH model standard errors using robust estimators and found that they made no difference in the significance of the values.

Some major points regarding the heterogeneous negative binomial are:

- Dispersion predictors (lnalpha) specify which predictors most influence the value of the dispersion parameter.
- An analyst can use NB-H to verify preconceived sources of overdispersion in the data.
- Robust standard errors should be used as the default.

TABLE 5.9. Stata: Squirrel Data

```
. use nuts
. center ntrees, prefix(s) standard
. center height, prefix(s) standard
. center cover, prefix(s) standard
. global xvars "sntrees sheight scover"
. glm cones $xvars if dbh<.6, fam(pois) nolog
. nbreg cones $xvars if dbh<.6, nolog
. gnbreg cones $xvars if dbh<.6, nolog lnalpha($xvars)
. gnbreg cones $xvars if dbh<.6, nolog lnalpha($xvars) vce(robust)
```

```
<results not displayed>
```

5.5 SUMMARY

The negative binomial model is commonly used by statisticians and analysts to model overdispersed Poisson data. Typically, negative binomial regression is used as a catch-all model for overdispersed count data. Many times – perhaps most of the time – we don't know the reason for overdispersion. We may have suspicions, such that there are an excessive number of zeros in the data, that the data are correlated, or even that there are no zero counts in the data. There are a host of reasons why data can be overdispersed. There are also many models to address the variety of overdispersion we find in study data. But if we have no clear idea as to what is causing overdispersion in a Poisson model, analysts generally turn to negative binomial regression.

Negative binomial regression is a two-parameter model that is estimated using either maximum likelihood estimation or by IRLS estimation as a member family of generalized linear models (GLMs). When the negative binomial is estimated as a GLM, the dispersion parameter is entered into the IRLS estimating algorithm as a constant. Some IRLS routines exist that estimate the dispersion parameter, α, using an outside maximum likelihood procedure, which then sends the result back into the IRLS fitting algorithm. R, Stata, SAS, and Limdep use this method for negative binomial estimation.

Two models were discussed that extend the basic negative binomial model – the three-parameter NB-P and the heterogeneous negative binomial, NB-H. The NB-P model parmeterizes the negative binomial variance exponent, allowing analysts to determine whether the data are better modeled using NB1 or NB2. The NB-H model allows analysts to determine which predictors contribute model extradispersion. Both are valuable diagnostic models, as well as alternative negative binomial models in their own right.

There are a wide number of enhanced negative binomial commands or functions that adjust the basic negative binomial algorithm to allow estimation of complex count data. We will address many of these in the following pages. However, we first must address a mixture count model that is not negative binomial but is rather a mixture of Poisson and inverse Gaussian distributions. It is one of the three basic types of count models we discuss in the book.

## CHAPTER 6

Poisson Inverse Gaussian Regression

SOME POINTS OF DISCUSSION

- Why hasn't the PIG model been widely used before this?
- What types of data are best modeled using a PIG regression?
- How do we model data with a very high initial peak and long right skew?
- How do we know whether a PIG is a better-fitted model than negative binomial or Poisson models?

6.1 POISSON INVERSE GAUSSIAN MODEL ASSUMPTIONS

The Poisson inverse Gaussian (PIG) model is similar to the negative binomial model in that both are mixture models. The negative binomial model is a *mixture of Poisson and gamma distributions*, whereas the inverse Gaussian model is a *mixture of Poisson and inverse Gaussian distributions*.

Those of you who are familiar with generalized linear models will notice that there are three GLM continuous distributions: normal, gamma, and inverse Gaussian. The normal distribution is typically parameterized to a log-normal distribution when associated with count models, presumably because the log link forces the distribution to have only nonnegative values. The

	POI	NB1	NB2	PIG	NB-P
TABLE 6.1. Mean–Variance Comparison					
Mean	μ	μ	μ	μ	μ
Variance	μ	$\mu + \alpha\mu$	$\mu + \alpha\mu^2$	$\mu + \alpha\mu^3$	$\mu + \alpha\mu^\rho$
Inv Variance	–	$\mu + \mu/\delta$	$\mu + \mu^2/\theta$	$\mu + \mu^3/\phi$	

Poisson and negative binomial (both NB2 and NB1) models have log links. Recall that the negative binomial is a mixture of the Poisson and gamma distributions, with variances of μ and μ^2/v, respectively. We inverted v so that there is a direct relationship between the mean, dispersion, and variance function. Likewise, the inverse Gaussian is a mixture of Poisson and inverse Gaussian distributions, with an inverse Gaussian variance of μ^3/ϕ. The mean–variance relationship of the Poisson, negative binomial, PIG, and NB-P, with accepted symbols for the inversely parameterized dispersion, is given in Table 6.1.

I parameterize the PIG distribution and associated regression model as the negative binomial is traditionally parameterized, with a direct relationship between the mean and dispersion parameter. Doing this directly relates the mean and amount of overdispersion in the data. The dispersion for the PIG model will also be given the name *alpha*, α. The dispersion is often given as *nu*, v, or *phi*, ϕ. I will use ϕ to refer to the dispersion parameterized indirectly with the mean such that $\phi = 1/\alpha$. This is similar to the relationship we discussed regarding the negative binomial, where *theta* is the inverse of α: $\theta = 1/\alpha$. We must make a clear distinction, though, in how θ is understood here. GLM theory symbolizes the exponential family link function as θ. R, however, has chosen to use the same symbol for the dispersion parameter as well. My caveat is to keep the distinction in mind when dealing with this class of models.

Recall Table 5.1, in which we compared the relationship of the mean and variance for several values of the Poisson and negative binomial distributions. The PIG variance can be added to the table for comparative purposes.

Table 6.2 shows us that with greater mean values of the PIG distribution come increasingly greater values of the variance. With a dispersion parameter greater than 1, greater values of the mean of the response variable in a PIG regression provide for adjustment of greater amounts of overdispersion than does the negative binomial model. Simply put, PIG models can better deal with highly overdispersed data than can negative binomial regression, particularly with data clumped heavily at 1 and 2.

TABLE 6.2. Mean Dispersion Variance Relationships							
μ	α	$\mu + \alpha\mu^2$	$\mu + \alpha\mu^3$	μ	α	$\mu + \alpha\mu^2$	$\mu + \alpha\mu^3$
.5	.5	0.625	.0563	5	.5	17.5	67.5
.5	1	0.75	.063	5	1	30	130
.5	2	1.0	.755	5	2	55	255
.5	5	1.75	1.125	5	5	130	380
1	5	1.5	1.5	10	.5	60	510
1	1	2.0	2.0	10	1	110	1010
1	2	3.0	3.0	10	2	210	2010
1	5	6.0	6.0	10	5	510	5010

μ: Poisson; $\mu + \alpha\mu^2$: NB2; $\mu + \alpha\mu^3$: PIG.

The foremost assumptions of the PIG model are similar to those underlying the negative binomial model. The key difference is the following:

PIG regression is used to model count data that have a high initial peak and that may be skewed to the far right as well as data that are highly Poisson overdispersed.

The PIG probability distribution, as a variety of Sichel distribution, can be given as

$$f(y; \mu, \alpha) = \sqrt{\frac{\phi}{2\pi y^3}} \exp\left(\frac{-\phi(y - \mu)^2}{2\mu^2 y}\right) \tag{6.1}$$

with $\{y, \mu, \phi\} > 0$.

Again, a value of the model rests in the fact that it can capture the modeling of count data that are sharply peaked and skewed to the far right (e.g., hospital data where most patients are discharged within the first few days and then the numbers taper off quickly, with a few perhaps lingering on for a long time). The negative binomial model may do a poor job of modeling such data, as will other count models. The PIG model is the only model we describe that can deal with this form of data.

The preceding PDF can be used to calculate PIG probabilities, as we did for the Poisson and NB2 distributions. Parameterizing $\alpha = 1/\phi$ for a direct relationship, with $\alpha = 1$, $y = 2$, and $\mu = .5$, we have a predicted PIG probability of

```
. di  sqrt(1/(1*2*_pi*2^3)) * exp((-(2-.5)^2)/(1* 2*.5^2 * 2))
.01486629
```

We can place this formula within a loop to obtain a range of probability values. Refer to Tables 1.2a and 1.2b for the PIG PDF formula using R and Stata, respectively. You can obtain the shape of the distribution for any mean and value of *alpha* you specify.

6.2 CONSTRUCTING AND INTERPRETING THE PIG MODEL

6.2.1 Software Considerations

No commercial statistical software offers the PIG model as a component of its official offerings. However, Stata has commands for its use, with the following three commands authored by Hardin and Hilbe:

pigreg: PIG regression
zipig: zero-inflated PIG (ZIPIG)
ztpig: zero-truncated PIG (ZTPIG)

The first two commands listed were written for the author's text Hardin and Hilbe (2012), whereas the ZTPIG model was written (by Hardin) for the present book. As we will later discover, zero-truncated count models are vital as the count component of hurdle models; we develop a PIG-logit hurdle model in the following chapter.

The R package **gamlss** (Rigby and Stasinopoulos 2008) supports PIG and ZIPIG modeling. It is the only R package to my knowledge that provides PIG modeling capability. I use both **pigreg** and **gamlss** in this chapter for example purposes. I am not aware of a SAS macro for PIG.

6.2.2 Examples

I begin by modeling the German health data using the **rwm1984** file that was used earlier in the book (see Table 6.3). The reduced model data will be used,

TABLE 6.3. R: Poisson Inverse Gaussian – rwm1984

```
library(gamlss); library(COUNT); library(msme); library(sandwich)
data(rwm5yr); rwm1984 <- subset(rwm5yr, year==1984)
summary(nbmod <- glm.nb(docvis ~ outwork + age, data=rwm1984))
vcovHC(nbmod)
sqrt(diag(vcovHC(nbmod, type="HC0")))
pigmod <- gamlss(docvis ~ outwork + age, data=rwm1984, family=PIG)
summary(pigmod)
exp(coef(pigmod))
vcovHC(pigmod)
sqrt(diag(vcovHC(pigmod, type="HC0")))
```

which includes *outwork* and *age* as predictors. Recall that *outwork* is binary and *age* is continuous, ranging from 25 to 64. Earlier, we centered the variable, which technically is the appropriate manner of handling a continuous predictor that does not start near 0 or 1. For pedagogical purposes, I'll leave *age* in its natural form. *docvis* ranges from 0 to 121, but there are only a few patients visiting a physician this much during the year. We will test the PIG model against NB2. Remember that the data have far more zero counts than allowed by the distributional assumptions:

```
. use rwm1984
NB2 MODEL
. glm docvis outwork age, nolog  vce(robust) fam(nb ml)

                    < not displayed >
. abic
AIC Statistic   =    4.303956         AIC*n     = 16673.525
BIC Statistic   =    4.304753         BIC(Stata) = 16698.572


. predict munb
(option n assumed; predicted number of events)


. linktest
_hatsq |   .0551027   .1963277    0.28   0.779   -.3296926    .439898
```

The negative binomial passes the linktest; it appears not to be misspecified:

```
PIG MODEL
. pigreg docvis outwork age, nolog vce(robust)

Poisson-Inverse Gaussian regression        Number of obs   =        3874
```

```
                                     Wald chi2(2)   =    1527.35
Log pseudolikelihood = -8378.5553    Prob > chi2    =     0.0000
-------------------------------------------------------------------
             |            Robust
      docvis |    Coef.   Std. Err.     z    P>|z|   [95% Conf. Interval]
-------------+-----------------------------------------------------
     outwork |  .5277197   .019414   27.18  0.000    .489669   .5657705
         age |  .0268643  .0010644   25.24  0.000   .0247781   .0289506
       _cons | -.2856271  .0556537   -5.13  0.000  -.3947063  -.1765479
-------------+-----------------------------------------------------
    /lnalpha |  1.344595  .0152366                  1.314732   1.374459
-------------+-----------------------------------------------------
       alpha |  3.836634  .0584573                  3.723753   3.952936
-------------------------------------------------------------------
. abich
AIC Statistic   =     4.327597      AIC*n      = 16765.111
BIC Statistic   =     4.328395      BIC(Stata) = 16790.158
```

You may compare the predicted counts for the NB2 (*munb*) and PIG (*mupig*) models as we have for other models:

```
. predict mupig
. linktest

_hatsq |   .3137388   .1355676    2.31   0.021    .0480311   .5794465

. pigreg , irr

Poisson-Inverse Gaussian regression     Number of obs   =       3874
                                        Wald chi2(2)    =    1527.35
Log pseudolikelihood = -8378.5553       Prob > chi2     =     0.0000
-------------------------------------------------------------------
             |            Robust
      docvis |     IRR    Std. Err.     z    P>|z|   [95% Conf. Interval]
-------------+-----------------------------------------------------
     outwork |  1.695063  .0329079   27.18  0.000   1.631776   1.760804
         age |  1.027228  .0010934   25.24  0.000   1.025088   1.029374
       _cons |  .7515428  .0418261   -5.13  0.000    .6738779   .8381587
-------------+-----------------------------------------------------
    /lnalpha |  1.344595  .0152366                  1.314732   1.374459
-------------+-----------------------------------------------------
       alpha |  3.836634  .0584573                  3.723753   3.952936
-------------------------------------------------------------------
```

The **linktest** evaluates whether the assumption of linearity has been violated. If the square of the *hat* matrix diagonal is significant, then the assumption has been violated – as is the case here. The test informs us that the PIG model is misspecified, that the PIG distribution is not the correct distribution to model these data. Note that the AIC and BIC statistics are greater than those of the negative binomial model.

We interpret the rate ratios the same way as for negative binomial regression, where patients who are not working were expected to see the doctor 70% more often in 1984 than working patients. The value of the PIG IRR for *age* is the same as we had for the negative binomial. Unfortunately, we cannot trust the estimates. They may be the same as for the negative binomial model, but the negative binomial model is itself highly overdispersed – in the sense that there is more variability in the data than allowed by the NB2 distributional assumptions.

I show the **gamlss** output to display the dispersion parameter, *sigma*. The value is 1.344. This value is on the log scale and must be exponentiated to obtain the model dispersion value. Note that the following display exponentiates the value of *sigma* to obtain 3.83435, the value of *alpha* resulting from the Stata estimation:

```
Mu Coefficients:
             Estimate  Std. Error  t value   Pr(>|t|)
(Intercept)  -0.28577     0.10801   -2.646   8.186e-03
outwork       0.52763     0.05563    9.485   4.096e-21
age           0.02686     0.00241   11.146   2.020e-28
--------------------------------------------------------------
Sigma link function:  log
Sigma Coefficients:
             Estimate  Std. Error  t value   Pr(>|t|)
(Intercept)     1.344     0.04166    32.26   1.792e-202

> exp(1.344)
[1] 3.83435
```

We next model the **medpar** data using a PIG model. Recall that the *type* categorical variable refers to the type of admission to a hospital. *Type1*, the reference level, is an elective admission, *type2* is urgent, and *type3* is emergency. We use a robust variance estimator since we found Poisson overdispersion in the data:

```
. pigreg los hmo white type2 type3, vce(robust) irr
Poisson-Inverse Gaussian regression          Number of obs   =       1495
                                              Wald chi2(4)    =    6047.35
Log pseudolikelihood = -4800.5418             Prob > chi2     =     0.0000
------------------------------------------------------------------------------
             |               Robust
         los |        IRR    Std. Err.      z    P>|z|    [95% Conf. Interval]
-------------+----------------------------------------------------------------
         hmo |   .9418869   .0055989   -10.07   0.000     .930977    .9529247
       white |     .87023   .0054969   -22.01   0.000    .8595228    .8810706
       type2 |   1.218342   .0060711    39.63   0.000    1.206501      1.2303
       type3 |   1.688646   .0149556    59.16   0.000    1.659586    1.718214
       _cons |   10.33001   .1612098   149.63   0.000    10.01883    10.65086
-------------+----------------------------------------------------------------
     /lnalpha |  -.6022063    .117711                    -.8329157   -.371497
-------------+----------------------------------------------------------------
       alpha |   .5476021   .0644588                     .4347797    .6897011
------------------------------------------------------------------------------
. abich
AIC Statistic   =    6.430156          AIC*n     = 9613.084
BIC Statistic   =    6.436512          BIC(Stata) = 9644.9424
AICH Statistic  =    6.430021          AICH*n    = 9595.1846

. linktest
------------------------------------------------------------------------------
         los |      Coef.   Std. Err.      z    P>|z|    [95% Conf. Interval]
-------------+----------------------------------------------------------------
        _hat |  -1.085516   3.213062    -0.34   0.735    -7.383002    5.211969
      _hatsq |   .4303325   .6625117     0.65   0.516    -.8681666    1.728832
       _cons |   2.505929   3.868191     0.65   0.517    -5.075586    10.08744
-------------+----------------------------------------------------------------
     /lnalpha |  -.6035841   .0571029                    -.7155038   -.4916645
-------------+----------------------------------------------------------------
       alpha |   .5468481   .0312266                     .4889457    .6116076
------------------------------------------------------------------------------
Likelihood-ratio test of alpha=0:   chibar2(01) = 4241.09
Prob>=chibar2 = 0.000
```

The AIC and BIC tests are statistically the same as for the negative binomial model. The linktest indicates that the model is not misspecified (_hatsq = .516). The negative binomial model linktest fails, though, with a hat-squared

TABLE 6.4. Interpretation of PIG Model Predictors

Private-pay patients are expected to stay in the hospital some 6% longer than HMO patients. $(1/.9419 = 1.06)$

Nonwhite patients are expected to stay in the hospital some 15% longer than white patients. $(1/.87 = 1.149)$

Urgent admissions are expected to stay in the hospital some 22% longer than elective admission patients. (1.218)

Emergency admissions are expected to stay in the hospital some 69% longer than elective admission patients. (1.689)

TABLE 6.5. R: Poisson Inverse Gaussian – medpar

```
library(gamlss); library(COUNT); library(msme); library(sandwich)
data(medpar)
rwm1984 <- subset(rwm5yr, year==1984)
summary(nbmod1 <- glm.nb(los ~ hmo + white + factor(type),data=medpar))
vcovHC(nbmod1)
sqrt(diag(vcovHC(nbmod1, type="HC0")))
pigmod1 <- gamlss(los ~ hmo + white + factor(type), data=medpar, family=PIG)
summary(pigmod1)
exp(coef(pigmod1))
vcovHC(pigmod1)
sqrt(diag(vcovHC(pigmod1, type="HC0")))
```

value of 0.036. It appears that the PIG model is to be preferred over the negative binomial model.

6.3 SUMMARY – COMPARING POISSON, NB, AND PIG MODELS

Two sets of data were modeled using Poisson, negative binomial, and Poisson inverse Gaussian regressions. Both data sets are Poisson overdispersed, as well as having more variation in them than allowed by negative binomial distributional assumptions.

The fact that the **rwm1984** data have an excessive number of zeros and **medpar** cannot have zero counts is likely the reason for the remaining overdispersion in the negative binomial model. The most reasonable test model for the **rwm1984** data is a zero-inflated model. The most reasonable

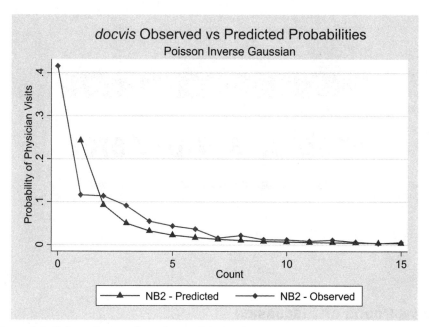

FIGURE 6.1. Poisson inverse Gaussian model: *docvis* observed vs. predicted probabilities.

test model for the **medpar** data is a zero-truncated model. We address both of these models in the following chapter.

I should mention that it is possible to program a Stata command like **countfit** to incorporate **pigreg** as an optional model. It does not exist at present. A graphic like Figure 5.4 for observed versus predicted PIG can nevertheless be made by substituting

```
qui gen 'newvar' = exp((-('i'-mu)^2)/('alpha'*2*'i'*mu^2)) *
(sqrt(1/('alpha'*2*_pi*'i'^3)))
```

in place of the negative binomial PDF in Table 5.3. Running the code produces Figure 6.1.

R code for the same figure may be created by substituting the R PIG PDF for the R code used to generate the negative binomial version of this figure. Graphics such as these are valuable ways in which an analyst can assess the fit of the model they believe best describes their study data.

## CHAPTER 7

Problems with Zeros

SOME POINTS OF DISCUSSION

- When is it necessary to use a zero-truncated count model?
- In what sense is a hurdle model called a two-part model and a zero-inflated model a mixture model?
- How can a count model be used as a binary component of a hurdle model?
- What is meant by "good" and "bad" zeros?
- How does an analyst decide between using a hurdle or zero-inflated model on data with excessive zeros?

The Poisson, negative binomial, and Poisson inverse Gaussian probability distributions all assume that the count data being modeled have zero counts. Exactly how many zeros is deterministically calculated based on the respective PDF and the mean of the counts in the distribution. For example, if we have a Poisson model with 100 observations and the mean of the count response is 3.0, we expect that the *percentage* of zero counts will be

```
. di exp(-3) * 3^0 / exp(lnfactorial(0))
.04978707
```

and that the *number* of zero counts will be .04978707 * 100 = 4.978707, or 5.

When there is a substantial disparity between the expected and observed zero counts in the data, given the mean of the response variable and number of observations in the model, a Poisson distribution should likely not be used to model the data. An adjustment can be made to the underlying PDF, or an entirely different model may be selected. The same is the case for negative binomial and PIG regressions.

In this chapter, we discuss what to do when there is no possibility of a count variable having a zero count, when there are far more zero counts than expected, when there are fewer zero counts than expected, and when the response can be broken up into two components, with one system generating zero counts and another having counts beginning with one (1). These are hurdle models:

```
R
=================================================
exp(-3) * 3^0 / exp(log(factorial(0)))
100* (exp(-3) * 3^0 / exp(log(factorial(0))))
=================================================
```

7.1 COUNTS WITHOUT ZEROS – ZERO-TRUNCATED MODELS

Many times we are asked to model count data that structurally exclude zero counts. Hospital length of stay data are an excellent example of count data that cannot have a zero count. When a patient first enters the hospital, the count begins. Upon registration, the length of stay is given as one (1). There can be no zero days, unless we are describing patients who do not enter the hospital, but this is an entirely different model. In such a case, there may be two generating processes – one generating 0's and another generating positive counts. I discuss this situation in the sections on hurdle models and zero-inflated count models.

When data structurally exclude zero counts, the underlying probability distribution must preclude this outcome in order to model the data properly. This is not to say that Poisson and negative binomial regressions are not commonly used to model such data; rather, the point is that in general they should not. The Poisson and negative binomial probability functions, and the other zero-truncated models described in this section, typically need to be amended to exclude zeros. At the same time, the amended probability function must provide for all of the constituent probabilities in the distribution to sum to 1. If the PDF does not sum to 1, it is no longer a probability distribution.

Zero-truncation models take two forms:

- as a model specifically designed to account for the absence of zero counts and
- as one of the options in constructing a left-truncated count model.

With respect to the second form, we will be addressing truncated models in Chapter 9. Truncating zeros is just one of the points in the distribution of counts where it may be cut. We will focus on the first meaning of zero truncation in this section, with the understanding that full-fledged truncated count models can easily be used for the purpose of cutting zeros from the distribution and adjusting the PDF accordingly.

We will discuss the following zero-truncated models in this section:

- zero-truncated Poisson (ZTP)
- zero-truncated negative binomial (ZTNB)
- zero-truncated Poisson inverse Gaussian (ZTPIG)
- zero-truncated generalized NB-P negative binomial (ZTNBP)
- zero-truncated Poisson log-normal (ZTPLN)

I should mention at this point that zero-truncated models have another use besides adjusting the underlying PDF of a model and the log-likelihood of its estimation algorithm so that they can model data that preclude the possibility of 0's. I am referring to their use as a component of hurdle models. We will discuss hurdle models in the following section.

Also to be mentioned is the fact that some statisticians have encouraged the use of *shifted Poisson* models when there are structurally no possible zero counts. That is, the counts are shifted left so that 1 counts become 0's, 2's become 1's, and so forth. A problem is that the resultant model is not based on data as they in fact exist. My caveat is that if a shifted model is employed on study data, it must be reported in study results.

7.1.1 Zero-Truncated Poisson (ZTP)

With respect to the Poisson distribution, the probability of a zero count is $\exp(-\mu)$, based on the PDF given in equation (1.3). This value needs to be subtracted from 1 and then the remaining probabilities rescaled on this difference. That is, the Poisson PDF is rescaled to exclude zero counts by dividing the PDF by $1 - \Pr(y = 0)$, which is $1 - \exp(-\mu)$. This in effect apportions the withdrawal of zero counts from each element of the Poisson PDF:

$$f(y; \mu) = \frac{e^{-\mu_i} \mu_i^{y_i}}{(1 - \exp(-\mu_i)) y_i!} \tag{7.1}$$

The resulting log-likelihood function, with $\mu = \exp(x\beta)$, is

$$\mathcal{L}(\mu; y | y > 0) = \sum_{i=1}^{n} \{ y_i (x_i'\beta) - \exp(x_i'\beta) - \ln \Gamma(y_i + 1)$$

$$- \ln[1 - \exp(-\exp(x_i'\beta))] \} \tag{7.2}$$

The model is no longer a GLM and cannot be estimated from within its umbrella. It is nearly always estimated using a full maximum likelihood algorithm. We therefore parameterize the log-likelihood in terms of *xb* and not μ.

Perhaps the most important thing to remember about zero truncation is that it is not really needed unless the mean of the distribution of counts to be modeled is low. Remember the Poisson PDF earlier in this chapter. We saw that, for a mean of 3, there should be only 5% 0's in the data. For a 100-observation data set, only 5 out of 100 counts should be 0's. However, for a mean of 5, we expect 2/3 of 1% of the counts to be 0's (i.e., about 7 out of 1000):

```
. di exp(-5) * 5^0 / exp(lnfactorial(0))
.00673795
```

For a mean of 12, we expect .0006% 0's, or 6 out of a million 0's:

```
. di  %10.9f exp(-12) * 12^0 / exp(lnfactorial(0))
0.000006144
```

If we have a model for which the count response variable has a mean greater than 5, we simply should not expect any 0 counts at all. So:

- For a Poisson model, if the mean is 5 or more, expect less than 1% 0's in the data set.

- Check the mean and number of observations of the data first before becoming concerned that you need a zero-truncated model.
- In general, employ robust standard errors for all zero-truncated models.

As an example, we use the **medpar** data. The response, *los*, has a mean of

```
. sum los
    Variable |        Obs        Mean    Std. Dev.        Min        Max
-------------+-------------------------------------------------------------
         los |       1495    9.854181    8.832906          1        116
```

With a mean of nearly 10 and with 1495 observations, we expect .079 0's and have none. Because of this, we expect there to be only minuscule differences between a standard and zero-truncated Poisson. If the mean had been 2, the differences would have been much more substantial. Finally, the ZT Poisson model that follows is not misspecified:

```
PREDICTED PROBABILITY OR 0 COUNTS FOR MEAN=9.854181
. di %10.9f exp(-9.854181) * 9.854181^0 / exp(lnfactorial(0))
0.000052527

PREDICTED 0 COUNTS FOR MEAN=9.854181
. di %10.9f exp(-9.854181) * 9.854181^0 / exp(lnfactorial(0)) * 1495
0.078528040
```

Comparing a Poisson and a zero-truncated Poisson on the **medpar** data, we have the following results. R code is provided in Table 7.1.

TABLE 7.1. R: Poisson and Zero-Truncated Poisson

```
library(msme)
library(gamlss.tr)
data(medpar)
poi <- glm(los~ white + hmo + factor(type), family=poisson, data-medpar)
summary(poi)
ztp <- gamlss(los~ white + hmo + factor(type),data=medpar, family=PO)
gen.trun(0, "POI", type="left", name = "lefttr")
lt0poi <- gamlss(los~white+hmo+ factor(type), data=medpar,
    family=POlefttr)
summary(lt0poi)
```

```
. glm los white hmo type2 type3, fam(poi) nolog nohead
POISSON
```

```
-----------------------------------------------------------------------
             |                OIM
       los |     Coef.    Std. Err.      z    P>|z|    [95% Conf. Interval]
-----------+-----------------------------------------------------------
     white |  -.153871    .0274128   -5.61   0.000   -.2075991   -.100143
       hmo |  -.0715493    .023944   -2.99   0.003   -.1184786    -.02462
     type2 |   .2216518   .0210519   10.53   0.000    .1803908   .2629127
     type3 |   .7094767    .026136   27.15   0.000    .6582512   .7607022
     _cons |   2.332933   .0272082   85.74   0.000    2.279606    2.38626
-----------------------------------------------------------------------
```

```
. abic
AIC Statistic   =    9.276131          AIC*n     =  13867.815
BIC Statistic   =    9.280208          BIC(Stata) = 13894.365
```

```
. ztp los white hmo type2 type3, nolog
...
ZERO-TRUNCATED POISSON
```

```
-----------------------------------------------------------------------
       los |     Coef.    Std. Err.      z    P>|z|    [95% Conf. Interval]
-----------+-----------------------------------------------------------
     white | -.1539437    .0274166   -5.61   0.000   -.2076792  -.1002081
       hmo | -.0716485    .0239636   -2.99   0.003   -.1186164  -.0246807
     type2 |  .2217806    .0210563   10.53   0.000     .180511   .2630502
     type3 |  .7096162    .0261385   27.15   0.000    .6583857   .7608467
     _cons |   2.33286    .0272121   85.73   0.000    2.279526   2.386195
-----------------------------------------------------------------------
```

```
. abic
AIC Statistic   =    9.275884          AIC*n     =  13867.447
BIC Statistic   =    9.279961          BIC(Stata) = 13893.996
```

There is in fact very little difference in the models. The guideline is that if the mean of the count response variable is high – perhaps over 4 – there is typically no need to model the data using a zero-truncated model.

7.1.2 Zero-Truncated Negative Binomial (ZTNB)

The logic of the zero-truncated negative binomial is the same as for the ZTP. The probability of a zero count is

$$(1 + \alpha\mu)^{-1/\alpha} \tag{7.3}$$

TABLE 7.2. R: Zero-Truncated Negative Binomial

```
library(msme); library(gamlss.tr)
data(medpar)
nb <- nbinomial(los~ white + hmo + factor(type), data=medpar)
summary(nb)
ztnb <- gamlss(los~ white + hmo + factor(type), data=medpar, family=NBI)
gen.trun(0, "NBI", type="left", name = "lefttr")
lt0nb <- gamlss(los~white+hmo+ factor(type), data=medpar, family=NBIlefttr)
summary(lt0nb)
```

Subtracting equation (7.3) from 1, logging the result and dividing the reparameterized negative binomial log-likelihood by the result, we have

$$\mathcal{L}(\mu; y | y > 0) = \sum_{i=1}^{n} \left\{ \mathcal{L}_{NB2} - \ln\left[1 - \left\{1 + \alpha \exp\left(x_i'\beta\right)\right\}^{-1/\alpha}\right]\right\} \quad (7.4)$$

where the first term in the braces is the standard NB2 log-likelihood function, given in equation (5.3). Note that since the log-likelihood is on the log scale, the term "1 Pr(0)" must also be logged; i.e., $\log(1 - Pr(0))$.

As an example, we use the same **medpar** data:

```
. ztnb los white hmo type2 type3, nolog
. . .
```

los	Coef.	Std. Err.	z	P>|z|	[95% Conf. Interval]	
white	-.1345583	.0757438	-1.78	0.076	-.2830134	.0138968
hmo	-.0726664	.0589298	-1.23	0.218	-.1881666	.0428339
type2	.2344362	.0559015	4.19	0.000	.1248713	.3440011
type3	.7355978	.084056	8.75	0.000	.5708511	.9003445
_cons	2.272518	.0752203	30.21	0.000	2.125089	2.419947
/lnalpha	-.6007157	.0549884			-.708491	-.4929403
alpha	.548419	.0301567			.4923866	.6108277

```
Likelihood-ratio test of alpha=0:  chibar2(01) = 4354.66
Prob>=chibar2 = 0.000

. abic
AIC Statistic  =   6.364409          AIC*n     = 9514.792
BIC Statistic  =   6.370764          BIC(Stata) = 9546.6514

. linktest
...
```

| _hatsq | .2424067 | .4024168 | 0.60 | 0.547 | -.5463157 | 1.031129 |

The AIC statistics for the ZTNB are substantially lower than those associated with the ZTP, indicating a much better fit.

A standard NB2 model is displayed for comparison. Note that the coefficients differ more than the ZTP did compared with the Poisson model. Note also that the AIC and BIC values are substantially lower for the zero-truncated model compared with the standard model. Recall that the values for the Poisson and ZTP were nearly identical. This is because more 0 counts were predicted for the negative binomial with the mean of *los*, and therefore more adjustment had to be made to adjust the truncated model PDF to sum to 1:

```
. glm los white hmo type2 type3, fam(nb ml) nolog nohead
-----------------------------------------------------------------------------
             |                 OIM
       los   |    Coef.   Std. Err.      z    P>|z|    [95% Conf. Interval]
---------+-------------------------------------------------------------------
     white   | -.1290654   .0685416   -1.88   0.060   -.2634046    .0052737
       hmo   | -.0679552   .0532613   -1.28   0.202   -.1723455    .0364351
     type2   |   .221249   .0505925    4.37   0.000    .1220894    .3204085
     type3   |  .7061588   .0761311    9.28   0.000    .5569446    .8553731
     _cons   |  2.310279   .0679472   34.00   0.000    2.177105    2.443453
-----------------------------------------------------------------------------

. abic
AIC Statistic   =   6.424718          AIC*n       = 9604.9531
BIC Statistic   =   6.428794          BIC(Stata)  = 9631.5029
```

The model is not misspecified, passing the linktest ($p = 0.547$). The likelihood ratio test tells us that the zero-truncated model is preferable to a standard negative binomial model.

Calculating the probability of zero counts for a negative binomial distribution is important when deciding if a negative binomial model would benefit by using a zero-truncated negative binomial. To determine the number of zeros that is expected for a model whose mean is 2 and alpha is 1, use the code in the following table and multiply the result by the number of observations in the model.

Stata: Calculate NB2 expected 0's for given α and μ

```
==============================================================
. gen a = 1
. gen mu = 2
. gen y=0
```

```
. di exp(y*log((a*mu)/(1+a*mu))-(1/a)*log(1+a*mu)+lngamma(y+1/a)-
> lngamma(y+1)-lngamma(1/a)) * <number of observations in model>
=========================================================
.33333333 * <number of observations in the model.
```

R: Calculate NB2 expected 0's for given α and μ

```
=========================================================
a <- 1; mu <- 2 ; y <- 0
exp(y*log(a*mu/(1+a*mu))-(1/a)*log(1+a*mu)+
log(gamma(y +1/a))-log(gamma(y+1))-log( gamma(1/a)))
=========================================================
[1] 0.3333333
```

To demonstrate that the sum of y probability values from 0 to 100 is equal to 1, we do the following:

R: Proof that sum of y probabilities from 0 to 100 is 1

```
====================================================
a <- 1 ; mu <- 2 ; y <- 0:100
ff <- exp(y*log(a*mu/(1+a*mu))-(1/a)*log(1+a*mu)+
   log(gamma(y +1/a))-log(gamma(y+1))-log( gamma(1/a)))
sum(ff)
====================================================
[1] 1
```

In the negative binomial model for the **medpar** data, the response, *los*, has a mean of 9.854181. Inserting that into the preceding formulae gives us a predicted probability of a zero count of 0.09213039. Multiplying by 1495 gives us 137.735 as the expected number of zero counts. So a sizable adjustment must have been made to the distribution since the zeros have been truncated from the model, as mentioned earlier.

7.1.3 Zero-Truncated Poisson Inverse Gaussian (ZTPIG)

The zero-truncated PIG is a new model with this book and will play an important role in hurdle models in the following section. I refer you to the previous chapter, devoted to this class of models, which aims to deal with data that are highly overdispersed.

The probability of a zero (0) PIG count is

$$\Pr(Y = 0) = \exp\left\{\frac{1}{\alpha}(1 - \sqrt{1 + 2/\alpha\mu})\right\} \tag{7.5}$$

where the dispersion statistic is given a direct relationship with the mean (i.e., $\alpha\mu$). Dividing the PIG PDF by $1 -$ equation (7.5) results in a zero-truncated PDF:

```
. ztpig los hmo white type2 type3, nolog

Zero-truncated PIG regression              Number of obs   =      1495
                                           LR chi2(4)      =     63.10
                                           Prob > chi2     =    0.0000
Log likelihood = -4778.6977                Pseudo R2       =    0.0066
-------------------------------------------------------------------------
       los |     Coef.   Std. Err.      z    P>|z|    [95% Conf. Interval]
-----------+-------------------------------------------------------------
       hmo | -.0643421    .0593155   -1.08   0.278   -.1805983    .0519142
     white | -.1478711    .0747265   -1.98   0.048   -.2943324   -.0014098
     type2 |  .2098073    .0554261    3.79   0.000    .1011741    .3184405
     type3 |  .5427813    .0793471    6.84   0.000    .3872638    .6982988
     _cons |  2.324182    .0741818   31.33   0.000    2.178788    2.469576
-----------+-------------------------------------------------------------
  /lnalpha | -.5109055    .0608871                    -.630242    -.391569
-----------+-------------------------------------------------------------
     alpha |  .5999521    .0365293                    .5324629    .6759954
-------------------------------------------------------------------------
Likelihood-ratio test of delta=0:   chibar2(01) = 4300.05
Prob>=chibar2 = 0.000


. abich
AIC Statistic   =    6.400933         AIC*n      = 9569.3955
BIC Statistic   =    6.407289         BIC(Stata) = 9601.2549
AICH Statistic  =    6.400798         AICH*n     = 9551.4971
```

The model is ideal for highly skewed data and highly overdispersed data that cannot have zero counts. No R code is currently available for this model. Note that a linktest failed to prove the model as misspecified. However, comparing AIC and BIC values, the model does not fit the data as well as the negative binomial.

Note: The **azsurgical** data on this books's web site is ideal for modeling with a ZTPIG. It is appendectomy data modeling *los* on *procedure*. The task is to discover the extent to which LOS relates to having open or laprascopic surgery.

7.1.4 Zero-Truncated NB-P (ZTNBP)

```
. ztnbp los hmo white type2 type3,

Zero-truncated Negbin-P model          Number of obs   =      1495
                                       Wald chi2(4)    =     50.61
Log likelihood = -4740.3872            Prob > chi2     =    0.0000
------------------------------------------------------------------
      los |    Coef.   Std. Err.      z    P>|z|   [95% Conf. Interval]
----------+-------------------------------------------------------
      hmo | -.0646661   .0483063   -1.34   0.181   -.1593447    .0300126
    white |  -.062624   .0656351   -0.95   0.340   -.1912665    .0660186
    type2 |  .2034899   .0507239    4.01   0.000    .1040729    .3029069
    type3 |  .7023193   .1081974    6.49   0.000    .4902563    .9143823
    _cons |  2.216729   .0650517   34.08   0.000     2.08923    2.344227
----------+-------------------------------------------------------
  /lnalpha | -3.279848   .7890374   -4.16   0.000   -4.826332   -1.733363
       /P |  3.177916   .3525702    9.01   0.000    2.486891    3.868941
----------+-------------------------------------------------------
    alpha |   .037634   .0296946    1.27   0.205    .0080159    .1766893

. abich
AIC Statistic   =     6.35102      AIC*n      = 9494.7744
BIC Statistic   =     6.359878     BIC(Stata) = 9531.9434
AICH Statistic  =     6.353812     AICH*n     = 9471.6875
```

This is the best-fitted model, if we are to believe the AIC and BIC statistics. We discussed this model in Chapter 4 as a three-parameter model aimed at advising the analyst whether NB2 or NB1 better fits the data. It appears that a PIG model is preferable to NB1 or NB2 since the NB-P power parameter is 3.18. It would be a mistake to interpret ρ in this manner, though, since NB-P is based on a negative binomial distribution, not a PIG. However, given that the ZTNBP model is better fitted than ZTPIG, it appears likely that there are other causes of overdispersion. Keep in mind that a model with excessive zero counts will almost always be overdispersed, but there may be other sources of overdispersion as well. The NB-P model may in effect be adjusting for multiple sources of overdispersion. Recall that, for the NB and PIG models, the dispersion statistic has a single value. If there is heterogeneity in the dispersion, then they are not suitable models to adjust for it. Heterogeneous NB and NB-P are suitable models, though, as are the three-parameter generalized negative binomial models we discuss in Chapter 9.

A linktest was given to the model, indicating that it is not misspecified. Recall that the ZTPIG model also passed the linktest and appears well specified.

7.1.5 Zero-Truncated Poisson Log-Normal (ZTPLN)

The probability function for the zero-truncated Poisson log-normal is discussed at length in Fabermacher (2011, 2012) in the context of using a hurdle model to predict the severity of automobile crashes. The model **ztprm**, authored by Fabermacher, uses an adaptive Gauss-Hermite quadrature method for model estimation. It is recommended to use a greater number of quadrature points to gain accuracy, but at the expense of time. The author of the software recommends that users check the sensitivity of the results:

```
. ztpnm los hmo white type2 type, vuong

Zero-truncated Poisson normal mixture model      Number of obs   =      1495
                                                 Wald chi2(4)    =     56.68
Log likelihood = -4770.7076                      Prob > chi2     =    0.0000
------------------------------------------------------------------------------
         los |     Coef.   Std. Err.      z    P>|z|    [95% Conf. Interval]
-------------+----------------------------------------------------------------
         hmo |  -.0604598   .0573828   -1.05   0.292    -.1729281    .0520084
       white |  -.1407904   .0736062   -1.91   0.056    -.2850559    .0034751
       type2 |   .2025758   .0544102    3.72   0.000     .0959339    .3092177
       type3 |   .5164477   .0828975    6.23   0.000     .3539716    .6789238
       _cons |   2.096475   .0727834   28.80   0.000     1.953822    2.239128
-------------+----------------------------------------------------------------
    /lnsigma |  -.3543547   .0263183  -13.46   0.000    -.4059376   -.3027719
-------------+----------------------------------------------------------------
       sigma |    .701626   .0184656   38.00   0.000     .6663518    .7387676
Vuong test of ztpnm vs. ztnb: z =    -3.19   Pr>z =  0.9993

. abich
AIC Statistic   =     6.390244        AIC*n        =   9553.415
BIC Statistic   =       6.3966        BIC(Stata)   =  9585.2744
AICH Statistic  =     6.390109        AICH*n       =  9535.5166
```

This model is also not misspecified. A Vuong test against the ZTNB is not good, though, telling us that the ZTNB is the preferred model. Given the differences in AIC and BIC statistics for each model, this is not surprising.

7.1.6 Zero-Truncated Model Summary

- Zero-truncated models are used
  1. *when the count data structurally exclude zero counts and*
  2. *for implementation in a hurdle model.*
- Zero-truncated models rarely need to be used when the response variable has a mean greater than 5. When they are employed in a hurdle model, this guideline is not relevant.

7.2 TWO-PART HURDLE MODELS

Hurdle models were first discussed as far back as 1971 (Cragg 1971), but Mullahy (1986) is generally given credit for their current popularity in modeling count data. The foremost use of a hurdle model is to deal with count response variables that have more – or fewer – zero counts than allowed by the distributional assumptions of the count model being used to understand the data. Two types of models are commonly used to handle excess zeros: hurdle models and zero-inflated models.

The essential idea of a hurdle model is to partition the model into two parts – first, a binary process generating positive counts (1) versus zero counts (0); second, a process generating only positive counts. The binary process is typically modeled using a binary model, and the positive count process is modeled using a zero-truncated count model. The most commonly used hurdle models are the Poisson-logit, NB2-logit, and NB2-probit models. Mullahy used logit and cloglog binary models with Poisson and geometric count models. A geometric model is a negative binomial with the dispersion parameter constrained to the value of 1. It has the distributional appearance of a negative exponential distribution, but with discrete values, not continuous values.

Again, the notion of "hurdle" comes from considering the data as being generated by a process that commences generating positive counts only after crossing a zero barrier, or hurdle. Until the hurdle is crossed, the process generates zeros. The nature of the hurdle is left unspecified but may simply be considered as the point where data consist of positive counts. In this sense, the hurdle is crossed if a count is greater than zero. Note that the hurdle does not have to be at 0; it may be at any higher value. However, for the binary component, values below the hurdle point are given the value of 0, and above the hurdle point they are given the value of 1. Most hurdle models, though, have the hurdle at crossing 0, which is the form we discuss in this section. In any case, the two processes are conjoined using the log-likelihood

$$\mathcal{L} = \ln(f(0)) + \{\ln[1 - f(0)] + \ln P(t)\} \tag{7.6}$$

where $f(0)$ represents the probability of a zero count, and $P(t)$ represents the probability of a positive count. Equation (7.6) reads that the hurdle model log-likelihood is the log of the probability of $y = 0$ plus the log of $y = 1$ plus the log of y being a positive count.

In the case of a logit model, the probability of zero is

$$f(0) = P(y = 0; x) = \frac{1}{1 + \exp(x_i'\beta_b)} \tag{7.7}$$

and $1 - f(0)$ is

$$\frac{\exp(x_i'\beta_b)}{1 + \exp(x_i'\beta_b)} \tag{7.8}$$

which is the probability of $y = 1$. When constructing log-likelihood functions for hurdle models, the binary component (e.g., the logit component) in equation (7.8) is combined with a zero-truncated count model, with the terms combined as in equation (7.6).

Before giving examples of hurdle models, I should say something about the availability of software. R users can use the hurdle function that is part of the **pscl** package on CRAN for a variety of hurdle models. Stata users must use the hurdle models I published over a decade ago, or simply construct the model by estimating its components separately. All of the major hurdle models are available and are on the book's web site. SAS has at least one user-created macro. See Morel and Neerchal (2012).

7.2.1 Poisson and Negative Binomial Logit Hurdle Models

Both components of a hurdle model may be combined in a single algorithm, with the log-likelihood formed as in equation (7.6). The two components of a hurdle model may be estimated separately, with summary statistics calculated as shown here.

As an example, we model *docvis* on *outwork* and *age* using the German health data, **rwm1984**.

TABLE 7.3. R: Poisson-Logit Hurdle

```
library(pscl); library(COUNT)
data(rwm5yr); rwm1984 <- subset(rwm5yr, year==1984)
hpl <- hurdle(docvis ~ outwork + age, dist="poisson", data=rwm1984,
                          zero.dist="binomial", link="logit")
summary(hpl); AIC(hpl)
```

The command **hplogit** is used for modeling. It is an author-written command:

```
. hplogit docvis outwork age, nolog

Poisson-Logit Hurdle Regression          Number of obs   =       3874
                                         Wald chi2(2)    =     155.20
Log likelihood = -12165.205              Prob > chi2     =     0.0000
-----------------------------------------------------------------------
             |    Coef.   Std. Err.     z    P>|z|   [95% Conf. Interval]
-------------+---------------------------------------------------------
logit        |
     outwork |  .5247121   .0718972   7.30   0.000    .3837962    .665628
         age |  .0263374   .0030821   8.55   0.000    .0202966   .0323782
       _cons | -.9935547   .1351158  -7.35   0.000   -1.258377  -.7287326
-------------+---------------------------------------------------------
poisson      |
     outwork |  .2134047   .0191964  11.12   0.000    .1757804    .251029
         age |  .0116587   .0008481  13.75   0.000    .0099965    .013321
       _cons |   1.04142   .0397268  26.21   0.000    .9635565   1.119283
-----------------------------------------------------------------------
AIC Statistic =      6.282

. abich
AIC Statistic   =      6.283534      AIC*n       = 24342.41
BIC Statistic   =      6.285986      BIC(Stata)  = 24379.982
AICH Statistic  =      6.281605      AICH*n      = 24328.146
```

Next, let's model the two-part hurdle model separately. We can use the following code guidelines (see Table 7.4):

- Create a variable *visit* equal to 1 if *docvis* > 0 and 0 otherwise.
- Model *visit* on predictors – the binary component of the hurdle model.

TABLE 7.4. R: Components to Poisson-Logit Hurdle

```
visit <- ifelse(rwm1984$docvis >0, 1, 0)
table(visit)
logis <- glm(visit ~ outwork + age, data=rwm1984,
                    family=binomial(link="logit"))
summary(logis)
library(pscl)
hpl2 <- hurdle(docvis ~ outwork + age, data=rwm1984,
      dist = "poisson", zero.dist="binomial", link="logit")
summary(hpl2)
logit <- glm(visit ~ outwork + age, data=rwm1984,
            family=binomial(link="logit"))
summary(logit)
```

- Model *docvis* > 0 on predictors using a zero-truncated Poisson.
- Compare the result with the hurdle model results.

First, for the binary component, we have the following logistic regression model:

```
. gen visit=docvis>0
. logit visit outwork age, nolog

Logistic regression                       Number of obs  =       3874
                                          LR chi2(2)     =     164.64
                                          Prob > chi2    =     0.0000
Log likelihood = -2547.8024               Pseudo R2      =     0.0313
-------------------------------------------------------------------------
      visit |   Coef.   Std. Err.    z     P>|z|    [95% Conf. Interval]
------------+------------------------------------------------------------
    outwork |  .5247119  .0718972   7.30   0.000    .383796    .6656278
        age |  .0263374  .0030821   8.55   0.000    .0202966   .0323782
      _cons | -.9935545  .1351158  -7.35   0.000   -1.258377  -.7287324
-------------------------------------------------------------------------
. abich
AIC Statistic  =    1.316883       AIC*n     = 5101.605
BIC Statistic  =    1.317036       BIC(Stata) = 5120.3911
AICH Statistic =    1.315514       AICH*n    = 5095.2573
```

It is the same as the hurdle model. Next the count component is modeled:

```
. ztp docvis outwork age if docvis>0, nolog

Zero-truncated Poisson regression          Number of obs   =      2263
                                           LR chi2(2)      =    443.77
                                           Prob > chi2     =    0.0000
Log likelihood = -9617.4025                Pseudo R2       =    0.0226
-----------------------------------------------------------------------
      docvis |     Coef.   Std. Err.      z    P>|z|   [95% Conf. Interval]
-------------+---------------------------------------------------------
     outwork |  .2134047   .0191964    11.12   0.000    .1757804    .251029
         age |  .0116587   .0008481    13.75   0.000    .0099965    .013321
       _cons |   1.04142   .0397268    26.21   0.000    .9635565   1.119283
-----------------------------------------------------------------------

. abich
AIC Statistic   =   8.502344         AIC*n       = 19240.805
BIC Statistic   =   8.502605         BIC(Stata)  = 19257.979
AICH Statistic  =   8.500219         AICH*n      = 19234.209
```

Since the components of a hurdle model are separate, the log-likelihood and AIC statistics of the hurdle model are the sum of values. The AICn of the count component is 19240.805 and the AICn of the logit is 5101.605. The sum is 24342.41. Note that the AICn value of the **hplogit** model here is the same value. Other more complex formulae can be used to calculate the statistic (Hilbe 2011), but it is easier to simply sum the values of the two components.

The interpretation is now considered. Each predictor is evaluated in terms of the contribution it makes to each respective model. For example, with respect to the Poisson-logit model, a positive significant coefficient in the Poisson frame indicates that the predictor increases the rate of physician visits in the same manner as any Poisson model is interpreted. A positive coefficient in the logit frame is interpreted in such a manner that a one-unit change in a coefficient decreases the odds of no visits to the doctor by $\exp(\beta)$. For example, the logistic coefficient in the hurdle model for *outwork* is 0.5247121; therefore the odds of no visits to the physician in 1984 are decreased by $\exp(.5247121) = 1.6899722$, or about 69%.

We know from previous analyses that **rwm1984** is overdispersed. It is therefore reasonable to model the data as an NB-logit hurdle. If there is reason to believe that the binary component is not asymmetric about .5, perhaps you should try an NB-complementary loglog hurdle model. I will not explore these models here, but you should know that they are available on the book's web site.

Look at the changes in the model coefficients and particularly in the standard errors and model p-values. Also check the AIC and BIC statistics to determine whether the negative binomial component has contributed to a significant reduction or elimination of overdispersion:

```
. hnblogit docvis outwork age, nolog

Negative Binomial-Logit  Hurdle Regression    Number of obs   =     3874
                                              Wald chi2(2)    =   155.20
Log likelihood = -8307.3248                   Prob > chi2     =   0.0000
------------------------------------------------------------------------
            |     Coef.   Std. Err.     z    P>|z|   [95% Conf. Interval]
------------+-----------------------------------------------------------
logit       |
    outwork |    .524712   .0718972    7.30  0.000    .3837961    .6656279
        age |   .0263374   .0030821    8.55  0.000    .0202966    .0323782
      _cons |  -.9935545   .1351158   -7.35  0.000   -1.258377   -.7287325
------------+-----------------------------------------------------------
negbinomial |
    outwork |   .2844499   .0621091    4.58  0.000    .1627184    .4061815
        age |   .0158616   .0027161    5.84  0.000    .0105382     .021185
      _cons |   .3522567   .1336561    2.64  0.008    .0902956    .6142178
------------+-----------------------------------------------------------
   /lnalpha |   .7514482   .0926998    8.11  0.000    .5697601    .9331364
------------+-----------------------------------------------------------
      alpha |   2.120068   .1965298                   1.767843    2.542471
------------------------------------------------------------------------
AIC Statistic =      4.290

. abich
AIC Statistic   =     4.292372        AIC*n       = 16628.65
BIC Statistic   =     4.295791        BIC(Stata)  = 16672.484
AICH Statistic  =     4.290557        AICH*n      = 16611.166
```

The information criterion tests (AIC and BIC) are significantly lower, indicating a better fit. We can determine whether robust standard errors are required for the count component, and we can parameterize the coefficients as an incidence rate ratio by using the options *irr* and *vce(robust)*.

```
. hnblogit docvis outwork age, nolog irr vce(robust)
       .      .      .
```

		Robust				
	IRR	Std. Err.	z	P>|z|	[95% Conf.	Interval]
logit						
outwork	1.689972	.1214699	7.30	0.000	1.467905	1.945634
age	1.026687	.0031672	8.54	0.000	1.020498	1.032914
_cons	.3702583	.0502134	-7.33	0.000	.2838357	.4829948
negbinomial						
outwork	1.329031	.1087	3.48	0.001	1.132182	1.560105
age	1.015988	.0033321	4.84	0.000	1.009478	1.02254
_cons	1.422274	.2124444	2.36	0.018	1.061303	1.906017
/lnalpha	.7514482	.1196007	6.28	0.000	.5170351	.9858614
alpha	2.120068	.2535617			1.677048	2.68012

Robust standard errors differ from the model standard errors for the count component; we therefore want to keep them in a model.

We have not yet discussed obtaining predicted values from a hurdle model. We can separate models and obtain predictions for each component, or predict for a specified component in a multicomponent model. The latter is the best course to take. The following Stata code provides predicted values for the negative binomial model. If eq(#2) is not specified, the default is the first logistic model component. We call the fitted or predicted value *mu*:

```
. hnblogit_p mu, eq(#2) irr
. l mu docvis outwork age in 1/5
```

	mu	docvis	outwork	age
1.	3.349409	1	0	54
2.	3.798543	0	1	44
3.	4.74305	0	1	58
4.	3.925132	7	0	64
5.	3.042121	6	1	30

TABLE 7.5. R: NB2-Logit Hurdle (Assume Model from 7.3 Loaded)

```
hnbl <- hurdle(docvis ~ outwork + age, dist="poisson", data=rwm1984,
                    zero.dist="binomial", link="logit")
summary(hnbl); AIC(hnbl)
alpha <- 1/hnbl$theta ; alpha
exp(coef(hnbl))
predhnbl <- hnbl$fitted.values
```

The value of *mu* for the negative binomial component can be checked by hand using Stata code (for the first two observations). Note how the coefficient and component levels are declared:

```
. di exp([#2]_b[outwork]*0 + [#2]_b[age]*54 + [#2]_b[_cons])
3.349409

. di exp([#2]_b[outwork]*1 + [#2]_b[age]*44 + [#2]_b[_cons])
3.7985435
```

Marginal effects may be obtained for hurdle models as well. See Hilbe (2011) for guidelines and example code on how to construct and interpret hurdle model marginal effects. Briefly, however, for both Stata and R (see Table 7.5) it is easiest to obtain separate marginal effects for each component. For example, given the NB2-logit model we just reviewed, to obtain marginal effects at means for the negative binomial component, we do as follows:

```
STATA CODE
. ztnb docvis i.outwork age if docvis>0
. margins, dydx(*)  atmeans noatlegend
```

and for the logit component

```
. gen visit  = docvis>0
. logit visit i.outwork age
. margins, dydx(*) atmeans noatlegend
```

If you prefer to use the **margins** command after a hurdle command (perhaps you don't have a zero), you still need to call the **margins** command twice, once for each component. The default display of marginal effects is the binary component (equation #1). Either type the same command as done previously or one for each component,

```
. margins, dydx(*) atmeans noatlegend predict(equation(#1))
. margins, dydx(*) atmeans noatlegend predict(equation(#2))
```

Note that if an older command is used that does not support the i. prefix for factor variables, use the **tabulate** command as

```
. tab outwork, gen(outwork)
```

and use *outwork1* in the model.

7.2.2 PIG-Logit and Poisson Log-Normal Hurdle Models

Poisson and negative binomial logit models have been in use for a number of years now. I have found no previous mention in the literature of a Poisson inverse Gaussian-logit hurdle model, or PIG-logit hurdle. In fact, the PIG has not been associated with any binary component in the construction of a hurdle model of which I am aware.

The PIG-logit hurdle model will likely be of use to those who have a substantial amount of variability in the model – more than is accounted for by simply adjusting the zero components in the model. I will use the same data as for the Poisson model, but realize that they are not ideal data for this model.

We continue to use the same logit model as before. Now, however, we use a zero-truncated PIG on the data, just as we did when constructing the two components of the hurdle model ourselves. Remember that it is mandatory to exclude zero counts from the estimation with *docvis* > 0, since the zero-truncated model assumes that no zero counts exist:

```
. ztpig docvis outwork age if docvis>0, nolog

Zero-truncated PIG regression              Number of obs   =      3874
                                           LR chi2(2)      =     73.90
                                           Prob > chi2     =    0.0000
Log likelihood = -5697.9235                Pseudo R2       =    0.0064
------------------------------------------------------------------------
      docvis |    Coef.   Std. Err.    z    P>|z|   [95% Conf. Interval]
-------------+----------------------------------------------------------
```

```
   outwork |   .2933067    .0604065     4.86    0.000     .1749122    .4117011
       age |   .0147178    .0026478     5.56    0.000     .0095282    .0199073
     _cons |   .6937237    .1221457     5.68    0.000     .4543225    .9331248
-----------+----------------------------------------------------------------
   /lnalpha |  .4436434    .0637793                       .3186383    .5686485
-----------+----------------------------------------------------------------
     alpha |  1.558375     .099392                       1.375254    1.765879
----------------------------------------------------------------------------
Likelihood-ratio test of alpha=0:  chibar2(01) = 7838.96
Prob>=chibar2 = 0.000

. abich
AIC Statistic   =    5.039261          AIC*n     = 11403.847
BIC Statistic   =    5.040627          BIC(Stata) = 11426.745
AICH Statistic  =    5.036855          AICH*n    = 11394.568
```

The PIG-logit hurdle can be constructed by placing the two components together. The AICn statistic can be calculated for the hurdle model as

```
. di 11403.847+5101.605
16505.452
```

and AIC as $16505.452/3874 = 4.261$. We calculated it for the NB-logit hurdle using the **hnblogit** command. The AIC statistic is 4.029 and AICn is 16619.46. The PIG-logit mode is definitely a better fit to the data than the Poisson-logit or NB-logit:

```
----------------------------------------------------------------------------
     visit |     Coef.   Std. Err.      z    P>|z|     [95% Conf. Interval]
-----------+----------------------------------------------------------------
   outwork |   .5247119    .0718972     7.30    0.000      .383796    .6656278
       age |   .0263374    .0030821     8.55    0.000     .0202966    .0323782
     _cons |  -.9935545    .1351158    -7.35    0.000    -1.258377   -.7287324
----------------------------------------------------------------------------

----------------------------------------------------------------------------
     docvis |    Coef.   Std. Err.      z    P>|z|     [95% Conf. Interval]
-----------+----------------------------------------------------------------
   outwork |   .2933067    .0604065     4.86    0.000     .1749122    .4117011
       age |   .0147178    .0026478     5.56    0.000     .0095282    .0199073
     _cons |   .6937237    .1221457     5.68    0.000     .4543225    .9331248
-----------+----------------------------------------------------------------
   /lnalpha |  .4436434    .0637793                       .3186383    .5686485
-----------+----------------------------------------------------------------
     alpha |  1.558375     .099392                       1.375254    1.765879
----------------------------------------------------------------------------
```

The Poisson log-normal hurdle may be calculated in the same manner. The preceding logit model will be the same model used here as well:

```
. ztpnm docvis outwork age if docvis>0,

Zero-truncated Poisson normal mixture model   Number of obs   =      2263
                                              Wald chi2(2)    =     73.22
Log likelihood = -5697.6664                   Prob > chi2     =    0.0000
-------------------------------------------------------------------------
      docvis |    Coef.   Std. Err.     z    P>|z|   [95% Conf. Interval]
-------------+-----------------------------------------------------------
     outwork |  .2398099   .0496063   4.83   0.000   .1425833   .3370365
         age |  .0117142   .0021539   5.44   0.000   .0074927   .0159358
       _cons |  .5241504    .099247   5.28   0.000   .3296299   .7186708
-------------+-----------------------------------------------------------
     /lnsigma | -.0650437    .022556  -2.88   0.004  -.1092526  -.0208348
-------------------------------------------------------------------------
       sigma |  .9370265   .0211355  44.33   0.000    .896504   .9793807

. abich
AIC Statistic   =    5.039033          AIC*n     = 11403.333
BIC Statistic   =    5.040399          BIC(Stata) = 11426.23
AICH Statistic  =    5.036628          AICH*n    = 11394.055

. di 11403.333+5101.605    /// AIC value
16504.938
```

The AICn value is nearly identical to that for the PIG-logit hurdle model. I also tested the generalized NB-P–logit hurdle, with the result that its AIC value (16629.17) is substantially higher than for the PIG or log-normal hurdle models.

7.2.3 PIG-Poisson Hurdle Model

Our last example is of a PIG-Poisson model. How can we do that? Which model is the binary model? It's the Poisson model. We can perform a right censored at 1 Poisson model. Here I use the **cpoissone** censored count model I published in the late 1990s. Since then, James Hardin and I have written a general censored count model command that can be used for left, right, or interval censoring. We have done the same for a truncated model, which is used in Chapter 9. This is a very powerful way to handle hurdle models, but the interpretation is a bit difficult:

```
STATA CODE
. gen rcen=1
. replace rcen=-1 if docvis>=1
. cpoissone docvis outwork age, censor(rcen) cright(1) nolog

Censored Poisson Regression              Number of obs  =     3874
                                         Wald chi2(2)   =   171.15
Log likelihood = -2545.2969              Prob > chi2    =   0.0000
------------------------------------------------------------------
    docvis |    Coef.   Std. Err.     z    P>|z|   [95% Conf. Interval]
-----------+------------------------------------------------------
   outwork |  .3418076  .0456266    7.49   0.000   .2523811    .4312342
       age |  .0178382  .0020181    8.84   0.000   .0138827    .0217937
     _cons | -1.043595  .0915753  -11.40   0.000  -1.223079   -.8641103
------------------------------------------------------------------
AIC Statistic =        1.316
AICn          =     5098.184
```

The overall PIG-Poisson hurdle AIC can be calculated just as for the previous hurdle models. The AICn for the right censored at 1 Poisson is 5098.184, and from before we know that the AICn of the zero-truncated PIG is 11403.847. The sum is

```
. di 5098.184+11403.847     /// AIC
16502.031
```

which is the lowest AICn value we have calculated. Perhaps a PIG-NB hurdle would be better, or a PIG-PIG hurdle. I will let you determine the result. What I hope that you learned from this section is that there are various ways to model count data.

We could have tried a wide variety of models, such as Poisson–lognormal–negative binomial hurdle, or any variety of binary model, such as logit, probit, complementary loglog, or loglog, or a right censored at 1 count model. The choice to be used is the one that best fits the data.

R's **pscl** package supports the Poisson and negative binomial hurdle models but no others. I hope to develop R functions that match the Stata commands used with this book.

7.3 ZERO-INFLATED MIXTURE MODELS

7.3.1 Overview and Guidelines

There are a variety of count data situations that can be modeled using Poisson regression, and if a model is Poisson overdispersed, as a negative binomial regression. Negative binomial models can themselves be extradispersed when, for example, there is more variability in the data than is allowed given the distributional assumptions of the negative binomial distribution. An easy first check for negative binomial extradispersion is the Pearson dispersion statistic provided in the model output of most statistical packages. Users of R's **glm** and **glm.nb** functions, however, must calculate the statistic themselves, or use the author-written **P__disp.r** function. R users may use the author's **nbinomial** function, which is part of the **msme** package, for negative binomial and heterogeneous negative binomial regression. The dispersion parameter, Pearson Chi2, and dispersion statistics are provided in default output.

When zero counts are not possible for the data being collected (e.g., hospital length-of-stay data), it is usually necessary to employ either a zero-truncated Poisson (ZTP) or zero-truncated negative binomial (ZTNB) if the mean of the response is low. On the other hand, if there are far more observed zero counts than expected given the mean of the distribution, analysts typically employ a hurdle model, a zero-inflated Poisson (ZIP), or a zero-inflated negative binomial (ZINB). We have discussed hurdle models in the previous section and address zero-inflated models in the present one.

Prior to commencing our look at zero-inflated models, it might be instructive to provide a few guidelines:

- There may be times when it is not desirable to use a zero-inflated model when the data have excessive zero counts. For example, it may be that a negative binomial or a hurdle model will be all that is needed to best understand such data – or some other model. To appropriately employ a zero-inflated model on data with excessive zero counts, the analyst should have a theory as to why there are a class of observations having both observed and expected zero counts.
- Zero-inflated models can be thought of as finite mixture models (i.e., where there are supposed two data-generating mechanisms, one generating 0's and one generating the full range of counts).

- There are two different types of 0's in the data – one generated by a binary component modeling 0's and one in the count model component of the mixture model. Binary 0's have been called "bad" 0's and count model 0's "good" 0's. This distinction is a mathematical fiction but can be made a meaningful interpretation of zero-inflated models.
- There is ongoing research and discussion concerning the appropriateness of various fit tests for zero-inflated models, including the boundary likelihood ratio test. The discussion relates to whether one zero-inflated model can be considered nested within another. We will use the tests, but know that there is discussion regarding their value in assessing fit.

7.3.2 Fit Tests for Zero-Inflated Models

I will use three primary tests to evaluate each of the zero-inflated models discussed in this section:

- *Boundary likelihood ratio test:* a test of one ZI model against another that is presumably nested in it. An example is ZINB to ZIP, where p-values less than 0.05 ($p < 0.05$) indicate that the ZINB model is preferable to ZIP.
- *Vuong test:* a nonnested test of a zero-inflated model against the noninflated model; for example, ZINB to NB. p-values less than 0.05 ($p < 0.05$) indicate that the ZINB model is preferable to NB. The test statistic is normally distributed $N(0,1)$, with large positive values favoring the ZINB model and large negative values favoring the NB model. If the p-value is not significant, then we cannot use the test to decide between ZINB or NB. However, the Vuong test is biased toward favoring the zero-inflated model, for which correction factors have been developed.
- *AIC/BIC tests:* be certain to check whether a standard noninflated model might better fit a zero-inflated model. In particular, it may be that a negative binomial, PIG, or NB-P model fits the data better than a zero-inflated Poisson or zero-inflated negative binomial. I will generally use the acronyms for the models from now on to save space.

7.3.3 Fitting Zero-Inflated Models

Recall that the hurdle model is a two-part model where each component can be modeled separately. The zero-inflated models cannot be separated in that manner. There are overlapping zeros, which are estimated by both the binary and count components.

The first component is the binary, which is usually modeled as a logit or probit. Unlike the hurdle model, though, in which the binary component has 1's consisting of all counts greater than 0 and 0's consisting of the 0 counts in the data, the zero-inflated models have a binary component with a value of 1 for all 0's in the data and 0 for all other counts greater than 0. The count component simply models all of the counts from zero to greater than zero. In this sense, the binary model is directly modeling the 0's in the data by revaluing them to 1's. The positive count values are the combined reference level for the binary model.

The logic of a zero-inflated model is that counts are estimated as

$$Pr(Y = 0) = Pr(Bin = 0) + (1 - Pr(Bin = 0)) * Pr(Count = 0)$$
$$Pr(Y >= 0) = (1 - Pr(Bin = 0)) + PDFcount \tag{7.9}$$

Since we will use the logit model for the binary component of zero-inflated models, I can give the logit equation for the probability of 0's as $1/(1 + \exp(xb))$:

$$Pr(Bin = 0) = \frac{1}{1 + \exp(x\beta)} \tag{7.10}$$

For a zero-inflated Poisson model with a logit binary component, we have the equation

$$(y = 0) = \log\left(\frac{1}{1 + \exp(-x\beta_b)} + \frac{\exp(-\exp(x\beta))}{1 + \exp(x\beta_b)}\right)$$

$$(y > 0) = \log\left(\frac{1}{1 + \exp(-x\beta_b)}\right) - \exp(x\beta) + y(x\beta) - \log\Gamma(y + 1) \tag{7.11}$$

where the b subscript indicates that $x\beta$ is a binary model component and without the subscript that $x\beta$ is from the count component, in this case Poisson.

7.3.4 Good and Bad Zeros

The discussion of good and bad zeros comes foremost from ecology, where interest rests in how we come to know what a zero count is. The logic can be used for other disciplines as well – in health care regarding patients, in insurance when dealing with insured clients, in transportation in counting accidents, and so forth.

The idea is that 0 counts represent errors. For example, when counting the number of birdcalls made by a type of bird, zero counts can occur because of

- birds being quiet during the time they were being recorded and
- those for which you didn't record the calls because you were sitting in the wrong place or were there at the wrong time to observe the calls. Perhaps you even failed to show up for the recording session.

In this case, the first reason for zero counts represents "good" counts and comes from the count component of the model. The second reason, because no one shows up to do the counting, or because someone had just shot at the birds and they flew away, or someone shot and killed them, represents "bad" zeros, which form the binary component.

There is really no way to tell where the lack of counts came from, but many ecologists have considered it important to make this type of distinction. I mention this here since you may read about it in the literature, especially if researchers from outside ecology start to use this distinction. See Zuur (2012) for a complete guide to this method.

7.3.5 Zero-Inflated Poisson (ZIP)

We use the same German health data as before, with the same predictors in each component. Ideally, we would include more predictors and would model the data using robust standard errors. For pedagogical reasons, we keep it simple:

```
. zipcv docvis outwork age, nolog inflate(outwork age)   vuong

Zero-inflated Poisson regression              Number of obs   =      3874
                                              Nonzero obs     =      2263
                                              Zero obs        =      1611
Inflation model = logit                       LR chi2(2)      =    443.03
Log likelihood  = -12165.53                   Prob > chi2     =    0.0000
------------------------------------------------------------------------
     docvis |    Coef.  Std. Err.      z    P>|z|   [95% Conf. Interval]
------------+-----------------------------------------------------------
docvis      |
    outwork |  .2132031  .0192091   11.10   0.000    .1755539    .2508522
        age |  .0116248  .0008472   13.72   0.000    .0099643    .0132852
      _cons |  1.043192  .0396523   26.31   0.000    .9654749    1.120909
------------+-----------------------------------------------------------
```

```
inflate   |
   outwork | -.5145616   .0724867   -7.10   0.000  -.6566328   -.3724904
       age | -.025714    .0031096   -8.27   0.000  -.0318087   -.0196192
     _cons | .9453518    .1365165    6.92   0.000   .6777844   1.212919
------------------------------------------------------------------------
Vuong test of zip vs. standard Poisson: z =   19.26   Pr>z = 0.0000
                                                      Pr<z = 1.0000
            with AIC (Akaike) correction: z =   19.24   Pr>z = 0.0000
                                                      Pr<z = 1.0000
           with BIC (Schwarz) correction: z =   19.19   Pr>z = 0.0000
                                                      Pr<z = 1.0000

. abic
AIC Statistic   =      6.2837        AIC*n      = 24343.055
BIC Statistic   =      6.286153      BIC(Stata) = 24380.627
```

The rate ratios for the count component of the preceding table of model statistics can be displayed as

```
. zip, irr

Zero-inflated Poisson regression          Number of obs   =     3874
                                          Nonzero obs     =     2263
                                          Zero obs        =     1611
Inflation model = logit                   LR chi2(2)      =   443.03
Log likelihood  = -12165.53               Prob > chi2     =   0.0000
------------------------------------------------------------------------
    docvis  |     IRR    Std. Err.    z    P>|z|   [95% Conf. Interval]
------------+-----------------------------------------------------------
docvis      |
   outwork  | 1.237636   .0237739   11.10  0.000   1.191906    1.28512
       age  | 1.011693   .0008571   13.72  0.000   1.010014   1.013374
     _cons  | 2.838262   .1125436   26.31  0.000   2.626034   3.067641
------------+-----------------------------------------------------------
inflate     |
   outwork  | -.5145616  .0724867   -7.10  0.000  -.6566328  -.3724904
       age  | -.025714   .0031096   -8.27  0.000  -.0318087  -.0196192
     _cons  | .9453518   .1365165    6.92  0.000   .6777844   1.212919
------------------------------------------------------------------------
Vuong test of zip vs. standard Poisson: z =   19.26   Pr>z = 0.0000
```

The Vuong test, displayed directly under the Stata model output, aims to test whether ZIP is preferable to the traditional Poisson model. A resulting *p*-value of under 0.05 provides evidence that the ZIP model is preferred (i.e., the number of zero counts exceeds Poisson distributional assumptions).

TABLE 7.6. R: ZIP

```
library(pscl); library(COUNT)
data(rwm5yr) ; rwm1984 <- subset(rwm5yr, year==1984)
poi <- glm(docvis ~ outwork + age, data=rwm1984, dist="poisson")
zip <- zeroinfl(docvis ~ outwork + age | outwork + age, data=rwm1984,
    dist="poisson")
summary(zip)
print(vuong(zip,poi))
exp(coef(zinp))
round(colSums(predict(zip, type="prob")[,1:17]))   # expected counts
rbind(obs=table(rwm1984$docvis)[1:18])             # observed counts
```

Two correction tests for the Vuong statistic are displayed by using the **zipcv** command (Desmarais and Harden 2013). The Vuong statistic is biased toward the zero-inflated model because the same data are used to estimate both the binary and count component parameters. AIC- and BIC-based correction factors adjust for the extra parameters in the zero-inflated component. The AIC correction factor tends to prefer the zero-inflated model when it is correct, but it does not always reject the zero-inflated model when it is not correct; the BIC correction factor tends to prefer the single-equation model when it is correct. Interpret the tests with care. I refer you to Desmarais and Harden (2013) for details. The correction factors are not yet available in R or SAS. For this model, the standard Vuong and both correction tests provide support for the ZIP model.

Next, calculate the predicted counts of the count component and predicted probability of the binary component, which we have specified as a binary logistic model. Recall that zero counts are modeled in both the binary and count components, resulting in a mixture of distributions. This mixture of zeros from both the binary and count model components must be accommodated when calculating the predicted counts.

The predicted probability of the binary logistic component of the ZIP model is calculated straightforwardly using the inverse logistic link, $1/(1 + \exp(-xb))$ or $\exp(xb)/(1 + \exp(xb))$:

```
STATA CODE
. predict prob  /* predicted count */
. gen xb_c = [docvis]_cons  + [docvis]_b[outwork] *outwork +
    [docvis]_b[age]*age
. gen xb_b = [inflate]_cons + [inflate]_b[outwork]*outwork +
    [inflate]_b[age]*age
. gen pr0 = 1/(1+exp(-xb_b))
```

The predicted counts for the count component, however, reflect the mixture of zeros. The Poisson inverse link, exp(xb), defines the predicted counts, but each count observation must be adjusted by 1 minus the predicted probability of being in the binary component:

```
. gen prcnt=exp(xb_c)*(1-pr0)
. su docvis prob prcnt pr0

    Variable |       Obs        Mean    Std. Dev.        Min         Max
-------------+---------------------------------------------------------
      docvis |      3874    3.162881    6.275955          0         121
        prob |      3874    3.160131    1.140611   1.612873     5.70035
       prcnt |      3874    3.160131    1.140611   1.612873    5.700351
         pr0 |      3874   .4116941    .0980708   .2288347    .5750543
```

Note that the AIC and BIC statistics are ~6.28. These values are substantially lower than the Poisson model AIC and BIC of 8.07 but much higher than the negative binomial value of 4.30. A negative binomial model of the data, regardless of the excessive zero counts, is better than either standard Poisson or ZIP models. Given that ZIP is preferable to Poisson, however, it is reasonable to assume that a ZINB model will be preferable to an NB model.

A table of expected counts can be given for the hurdle model, which can be compared with the observed counts:

```
R CODE
> pred <- round(colSums(predict(zip, type="prob") [,1:13]))
> obs <- table(rwm1984$docvis)[1:13]
> rbind(obs, pred)

        0    1   2   3   4   5   6   7   8   9 10 11 12
obs  1611 448 440 353 213 168 141  60  85  47 45 33 43
pred 1611  75 177 285 355 364 321 250 176 114 68 38 20
```

7.3.6 Zero-Inflated Negative Binomial (ZINB)

```
. zinb docvis outwork age, nolog inflate(outwork age) zip vuong irr

Zero-inflated negative binomial regression    Number of obs   =     3874
                                              Nonzero obs     =     2263
                                              Zero obs        =     1611
```

```
Inflation model = logit                    LR chi2(2)      =     108.45
Log likelihood  = -8316.853                Prob > chi2     =     0.0000
------------------------------------------------------------------------
      docvis |      IRR    Std. Err.     z    P>|z|   [95% Conf. Interval]
-------------+----------------------------------------------------------
docvis       |
     outwork |  1.334501   .0839373    4.59   0.000    1.179723   1.509585
         age |  1.018774   .0028387    6.68   0.000    1.013226   1.024353
       _cons |  1.356756   .1847742    2.24   0.025    1.038911   1.771844
-------------+----------------------------------------------------------
inflate      |
     outwork | -1.312531   .6455665   -2.03   0.042   -2.577818  -.0472438
         age | -.0318529   .0119638   -2.66   0.008   -.0553015  -.0084043
       _cons | -.2309631   .5240838   -0.44   0.659   -1.258149   .7962223
-------------+----------------------------------------------------------
    /lnalpha |  .5832367   .0677161    8.61   0.000    .4505156   .7159578
-------------+----------------------------------------------------------
       alpha |  1.791829   .1213356                    1.569121   2.046146
------------------------------------------------------------------------
Likelihood-ratio test of alpha=0: chibar2(01) =   7697.35
                                        Pr>=chibar2 = 0.0000
Vuong test of zinb vs. standard negative binomial: z =      2.61
                                         Pr>z = 0.0046 <=
                                         Pr<z = 0.9954
       with AIC (Akaike) correction: z =      2.11    Pr>z = 0.0172
                                         Pr<z = 0.9828
      with BIC (Schwarz) correction: z =      0.58  Pr>z = 0.2824
                                         Pr<z = 0.7176

. abic
AIC Statistic   =    4.297291        AIC*n      = 16647.707
BIC Statistic   =    4.30071         BIC(Stata) = 16691.541

. predict prob  /* predicted count */
. gen xb_c = [docvis]_cons  + [docvis]_b[outwork] *outwork +
  [docvis]_b[age]*age
. gen xb_b = [inflate]_cons + [inflate]_b[outwork]*outwork +
  [inflate]_b[age]*age
. gen pr0 = 1/(1+exp(-xb_b))
```

The predicted counts for the count component reflect the mixture of zeros. The negative binomial inverse link, $\exp(xb)$, defines the predicted counts, but each count observation must be adjusted by 1 minus the predicted probability

TABLE 7.7. R: ZINB

```
library(pscl); library(COUNT)
data(rwm5yr); rwm1984 <- subset(rwm5yr, year==1984)
nb2 <- glm.nb(docvis ~ outwork + age, data=rwm1984)
zinb <- zeroinfl(docvis ~ outwork + age | outwork + age, data=rwm1984,
    dist="negbin")
summary(zinb)
print(vuong(zinb,nb2))
exp(coef(zinb))
pred <- round(colSums(predict(zinb, type="prob")[,1:13]))
                                                    # expected counts
obs <- table(rwm1984$docvis)[1:13]                  # observed counts
rbind(obs, pred)
```

of being in the binary component:

```
. gen prcnt=exp(xb_c)*(1-pr0)
. su docvis prob prcnt pr0
```

Variable	Obs	Mean	Std. Dev.	Min	Max
docvis	3874	3.162881	6.275955	0	121
prob	3874	3.161081	1.162163	1.590592	5.792892
prcnt	3874	3.161082	1.162163	1.590592	5.792892
pr0	3874	.1299529	.073266	.0270649	.2636108

R code is given in Table 7.7, and partial output follows the table:

```
R OUTPUT
Count model coefficients (negbin with log link):
            Estimate Std. Error z value Pr(>|z|)
(Intercept)  0.305096   0.136207   2.240    0.0251 *
outwork      0.288557   0.062889   4.588  4.47e-06 ***
age          0.018600   0.002786   6.676  2.45e-11 ***
Log(theta)  -0.583239   0.067736  -8.610   < 2e-16 ***

Zero-inflation model coefficients (binomial with logit link):
            Estimate Std. Error z value Pr(>|z|)
(Intercept) -0.23096    0.52445   -0.440   0.65966
outwork     -1.31252    0.64614   -2.031   0.04222 *
age         -0.03185    0.01196   -2.664   0.00772 **
---
```

```
Signif. codes:  0 '***' 0.001 '**' 0.01 '*' 0.05 '.' 0.1 ' ' 1

Theta = 0.5581
Number of iterations in BFGS optimization: 26
Log-likelihood: -8317 on 7 Df

> print(vuong(zinb,nb2))
Vuong Non-Nested Hypothesis Test-Statistic: 2.60572
(test-statistic is asymptotically distributed N(0,1) under the
null that the models are indistinguishable)
in this case:
model1 > model2, with p-value 0.00458407

> exp(coef(zinb))
count_(Intercept)      count_outwork          count_age  zero_(Intercept)
        1.3567548          1.3345011          1.0187743         0.7937695
     zero_outwork           zero_age
        0.2691399          0.9686489

OBSERVED AND EXPECTED OR PREDICTED ZINB COUNTS
> rbind(obs, pred)
         0   1   2   3   4   5   6   7  8  9 10 11 12
obs   1611 448 440 353 213 168 141  60 85 47 45 33 43
pred  1625 535 357 261 199 156 124 100 82 67 56 46 39
```

ZINB models may be compared with ZIP models using the boundary likelihood test, with *p*-values less than 0.05 indicating that the ZINB model is preferable to ZIP. ZINB models may be compared with NB (NB2) models by using the Vuong test. *p*-values less than 0.05 indicate that the ZINB model is preferable to NB. The test statistic is normally distributed, with large positive values favoring the ZINB model and large negative values favoring the NB model. An insignificant test result favors neither model.

The likelihood ratio test assesses the relationship of ZINB to ZIP. Here the test clearly prefers ZINB. The uncorrected Vuong and AIC-corrected tests show that the ZINB model is preferable to the standard negative binomial model ($p = 0.0046$ and $p = 0.0172$, respectively). The BIC-corrected Vuong prefers neither ($p = 0.2824$). The standard negative binomial AIC and BIC statistics are 16673.525 and 16698.572, which gives a 26-point drop in the AIC statistic for the ZINB model compared with the standard NB model, but only a 7-point drop in the BIC statistic. Given the AIC-corrected Vuong statistic preferring the ZI model and the BIC-corrected statistic preferring neither, it is likely reasonable to give an overall slight preference to ZINB. The difference in test results should be acknowledged in a study report.

Note that the binomial (logistic) component that models the zero counts does not exponentiate the coefficients, generating odds ratios. I recommend that this be done. The standard errors of the odds ratios are determined by multiplying each respective odds ratio by its associated model SE. Confidence intervals are exponentiated, as are the coefficients.

7.3.7 Zero-Inflated Poisson Inverse Gaussian (ZIPIG)

If the counts are distributed such that there are many counts in the lower range of numbers, with a long right skew, then a *Poisson inverse Gaussian* model may be preferable. From a look at the tabulated count values, this may indeed be the case:

```
. zipig docvis outwork age, nolog inflate(outwork age) vuong zip irr

Zero-inflated Poisson inverse gaussian regression
                                          Number of obs   =     3874
                                          Nonzero obs     =     2263
                                          Zero obs        =     1611
Inflation model = logit                   LR chi2(2)      =    83.16
Log likelihood  = -8246.001               Prob > chi2     =   0.0000
------------------------------------------------------------------------
      docvis |      IRR  Std. Err.     z   P>|z|   [95% Conf. Interval]
-------------+----------------------------------------------------------
docvis       |
     outwork | 1.343885  .0824651   4.82  0.000   1.191597    1.515635
         age | 1.015202  .0026721   5.73  0.000   1.009978    1.020453
       _cons | 1.965736  .2368998   5.61  0.000   1.552183    2.489472
-------------+----------------------------------------------------------
inflate      |
     outwork | -.5030342 .1153151  -4.36  0.000   -.7290476   -.2770207
         age | -.024547  .0047009  -5.22  0.000   -.0337607   -.0153334
       _cons | .3784043  .2047753   1.85  0.065   -.0229479    .7797565
-------------+----------------------------------------------------------
    /lnalpha | .4433829  .0638008   6.95  0.000   .3183356     .5684302
-------------+----------------------------------------------------------
       alpha | 1.557969  .0993997                 1.374838     1.765493
------------------------------------------------------------------------
Likelihood-ratio test of alpha=0: chibar2(01) =   7839.05
                                   Pr>= chibar2 =   0.0000
Vuong test of zipig vs. poisson inverse gaussian:
                    z =   8.81  Pr>z = 0.0000
```

```
         Bias-corrected (AIC) Vuong test:    z =   8.61  Pr>z = 0.0000
         Bias-corrected (BIC) Vuong test:    z =   7.99  Pr>z = 0.0000

. abic
AIC Statistic   =    4.260713           AIC*n       = 16506.002
BIC Statistic   =    4.264132           BIC(Stata)  = 16549.836
```

Predicted counts and zeros are obtained by

```
. predict prpc, n
. predict prp0, pr
. su docvis prpc prp0

    Variable |       Obs        Mean    Std. Dev.        Min         Max
-------------+-----------------------------------------------------------
      docvis |      3874    3.162881    6.275955          0         121
        prpc |      3874    3.167142    1.266035    1.37824    6.192106
        prp0 |      3874    .2987763    .0992726   .1334178     .489595

R
=========================================================
library(gamlss) ; data(rwm1984); attach(rwm1984)
summary(zpig <- gamlss(docvis ~ outwork + age, nu.fo= ~ outwork + age,
    family=ZIPIG, data=rwm1984))
# code for calculating vuong, LR test, etc. on book's web site
=========================================================
```

The AIC and BIC values are some 140 points lower than that of the zero-inflated negative binomial (ZINB), and 50 points lower than the zero-inflated generalized Poisson (ZIGP) (examined in the following chapter). Moreover, the Vuong test favors the ZIPIG model over the PIG. The test statistic is normally distributed, with large positive values favoring the ZIPIG model and large negative values favoring the PIG model. The likelihood ratio test is also significant, comparing the ZIPIG with the zero-inflated Poisson (ZIP).

The test clearly prefers ZIPIG to any of the models we have thus far used. To be sure, the excess of zero counts significantly impacts model overdispersion.

7.4 SUMMARY – FINDING THE OPTIMAL MODEL

We have gone through a number of different zero-truncated, zero-inflated, and hurdle models. Zero-truncated models are used primarily when the count response variable structurally has no capability of having zero counts and where the mean of the distribution is low (e.g., under 4 or 5). Zero-truncated models are also used in the construction of hurdle models. Specialized software for zero truncation may be used, or a general truncation command or function that is set at a truncation at 0 may be used for the truncated at zero component.

Hurdle models are used for count data that have more or fewer zeros than are assumed on the basis of the distribution of the count component. The standard hurdle model consists of a binary component and a count component, which may be estimated separately or as a single algorithm. However, the fitting algorithm of a hurdle model typically estimates both components separately, combining them at the point of displaying results to the screen. The idea is that there is first a binary data-generating mechanism that produces 0 counts until some transition stage occurs, which starts an event mechanism generating positive counts. That is, when the binary transition "1" occurs, counts are generated using a zero-truncated count model.

The binary component is usually constructed as a binary logit or probit model with 1 equal to all nonzero values of the response variable and 0 otherwise, and a zero-truncated count model. However, the count component may also be formed by a right censored at 1 count model. I showed an example of setting up such a model with a PIG-Poisson hurdle. It was a substantially better fit than the other hurdle models we considered for the example data.

A double hurdle model may be constructed from a combination of truncated models; for example, a double Poisson-logit hurdle model will have a 1/0 logit model as a standard hurdle, but then an interval truncated model with cut points at 0 and at the next hurdle point, and then a left-truncated model from the second hurdle point to the highest count. This can be done for truncated models with the software we address in Chapter 9.

I should also mention that a hurdle model is an appropriate model to use when the count response variable has either fewer or greater zero counts than are expected based on the distributional assumptions of the model.

Finally, a zero-inflated model is a mixture model that may also be referred to as a type of finite mixture model. A zero-inflated model is used when there are excessive zero counts in the data, and generally when one has a theory for why there are so many zeros. When using them, care must be taken to make sure that no other models exist that fit as well or better. The standard zero-inflated models include the Poisson and negative binomial models. We have discussed several other zero-inflated models (e.g., ZI PIG,

	AICn	BIC	AICHn	Model
TABLE 7.8. Comparative Fit of rwm1984 Data				
ZIP	24343.055	24380.627	24328.791	ZI Poisson
ZINB	16647.707	16691.541	16630.223	ZI negative binomial
NB-L hurdle	16628.650	16672.484	16611.166	negative binomial-logit hurdle
ZINBP	16625.041	16675.139	16603.963	ZI NB-P
ZIHNB	16624.355	16680.713	16599.256	ZI heterogeneous negative binomial
ZIGP	16559.494	16603.328	16542.010	ZI generalized Poisson (from Chap. 8)
ZIPIG	16506.002	16549.836	16488.520	ZI Poisson inverse Gaussian (PIG)
Pig-P hurdle	16502.031	16547.136	16489.821	PIG-Poisson hurdle

ZI generalized Poisson, ZI NB-P, and ZI heterogeneous NB). A zero-inflated generalized Waring regression is discussed in Chapter 9.

So, when do we model data with an excess of 0 counts with a hurdle model or a zero-inflated model? If there is a real separation of mechanisms producing the 0's and the positive counts, a hurdle model appears best. If there is an overlap of 0 values, then a zero-inflated model may be the best choice. Of course, the model having substantially lower information test statistics should be preferred, other considerations being equal. Keep in mind that a hurdle model is a two-part model, and a zero-inflated model a mixture of models.

It should also be mentioned that of all the zero-inflated models we have tried with the **rwm1984** data, the zero-inflated PIG appears to be the best-fitted model (see Table 7.8). A PIG-Poisson hurdle model, however, has slightly lower AIC and BIC values, but not enough to decide that it is a better-fitted model. The decision of which model to select should be based on context and how the models are to be interpreted.

## CHAPTER 8

Modeling Underdispersed Count Data – Generalized Poisson

SOME POINTS OF DISCUSSION

- How does a generalized Poisson model underdisperse?
- What is the relationship between the Poisson and generalized Poisson models?
- Is the generalized Poisson a mixture model? If so, what are the component distributions? How can a count model be used as a binary component of a hurdle model?

An analyst rarely encounters underdispersed Poisson data when dealing with real data. But it does happen, and several journal articles have been authored that use one method or another to provide parameter estimates for underdispersed data.

The problem with not paying attention to data that are underdispersed and modeling them by normal means is that the standard errors of the resulting model are overestimated. This leads to thinking that predictors are not significant when in fact they are. Normally, standard errors are overdispersed,

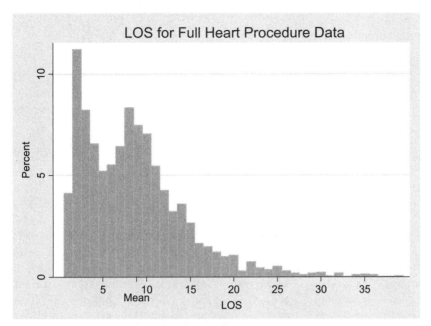

FIGURE 8.1. LOS for full heart procedure data.

leading an analyst to believe that predictors are significant when in fact they are not.

The generalized Poisson probability function is based on Consul (1989) and Famoye (1993), and this parameterization on Harris, Yang, and Hardin (2012):

$$f(y; \theta, \delta) = \frac{\theta_i \, (\theta_i + \delta y_i)^{y_i - 1} \, e^{-\theta_i - \delta y_i}}{y_i!}, \, y_i = 0, 1, 2, \dots \quad (8.1)$$

The types of count data that are underdispersed consist of data that are lumped more tightly together than should be expected based on Poisson and negative binomial distributional assumptions. Consider the **azprocedure** data that is on the book's web site. This data set is a record of how long patients were in the hospital following either a CABG (1) or PTCA (percutaneous transluminal coronary angioplasty) (0) *procedure*. Other predictors are *sex* (1 = male; 0 = female) and *admit* (1 = emergency/urgent; 0 = elective). We have found that it is rather severely overdispersed at 3.23, with a mean length of hospital stay of 8.2 days:

```
. sum los

    Variable |       Obs        Mean    Std. Dev.        Min         Max
-------------+-----------------------------------------------------------
         los |      3589    8.830872    6.926239          1          83

. hist los if los<40, title("LOS for full Heart Procedure Data") discrete
xlab( 5 8.83 "mean" 10 15 20 25 30 35) percent
POISSON REGRESSION
. glm los procedure sex admit, nolog fam(poi) vce(robust) nohead nolog eform
-------------------------------------------------------------------------
             |               Robust
         los |       IRR   Std. Err.      z    P>|z|    [95% Conf. Interval]
-------------+-----------------------------------------------------------
   procedure |  2.604873   .0567982   43.91   0.000    2.495896    2.718608
         sex |  .8779003   .0195083   -5.86   0.000    .8404854    .9169806
       admit |  1.395248   .0297964   15.60   0.000    1.338054    1.454887
       _cons |  4.443334   .1174981   56.40   0.000    4.218908    4.679698
-------------------------------------------------------------------------

. abich
AIC Statistic   =    6.264283         AIC*n     = 22482.514
BIC Statistic   =    6.265144         BIC(Stata) = 22507.256

. di e(dispers_p)   // Dispersion
3.2323804

. di e(N)             // Number of observations in model
3589
```

I'll next drop all lengths of stay greater than 8 days, which amounts to 1607 patients. That leaves 982 remaining, with a new mean LOS of 4.46 days. The dispersion has dropped to 0.79, indicating underdispersion:

```
. drop if los>8
. sum los

    Variable |       Obs        Mean    Std. Dev.        Min         Max
-------------+-----------------------------------------------------------
         los |      1982    4.465691     2.30982          1           8
```

A Poisson model on the data clearly demonstrates that the data are underdispersed:

```
. glm los procedure sex admit, nolog fam(poi) vce(robust) eform

Generalized linear models                No. of obs      =      1982
Optimization     : ML                    Residual df     =      1978
                                         Scale parameter =         1
Deviance       =  1574.440006            (1/df) Deviance = .7959757
Pearson        =  1562.747889            (1/df) Pearson  = .7900647
                                         AIC             =  4.020972
Log pseudolikelihood = -3980.783619      BIC             = -13442.26
------------------------------------------------------------------------
             |             Robust
      los    |    IRR    Std. Err.     z     P>|z|    [95% Conf. Interval]
----------+-------------------------------------------------------------
procedure  | 2.077086   .0344479   44.07   0.000    2.010655   2.145712
      sex  |  .9333992  .0185912   -3.46   0.001     .8976632   .9705579
    admit  | 1.363049   .0249235   16.94   0.000    1.315064   1.412784
    _cons  | 3.276827   .0690439   56.33   0.000     3.14426   3.414984
------------------------------------------------------------------------
. abich
AIC Statistic  =    4.020972         AIC*n      = 7969.5674
BIC Statistic  =    4.022532         BIC(Stata) = 7991.9346
```

Now, let's see if the generalized Poisson model picks up on the underdispersion. Remember, a negative binomial model cannot model data that are underdispersed. If it does succeed in estimating parameters, they are at the values of a Poisson model:

```
. gpoisson los procedure sex admit, nolog  vce(robust)

Generalized Poisson regression           Number of obs   =      1982
                                         Wald chi2(3)    =   1891.88
Prob > chi2                                              =    0.0000
Dispersion     = -.1195314               Prob > chi2     =    0.0000
Log pseudolikelihood = -3956.3794        Pseudo R2       =    0.1052
------------------------------------------------------------------------
             |             Robust
      los    |    IRR    Std. Err.     z     P>|z|    [95% Conf. Interval]
----------+-------------------------------------------------------------
procedure  | 2.051253   .0343166   42.94   0.000    1.985085   2.119628
      sex  |  .934638   .0186265   -3.39   0.001     .8988346   .9718675
    admit  | 1.366056   .0255772   16.66   0.000    1.316835   1.417118
    _cons  | 3.281648   .0706452   55.20   0.000    3.146066   3.423072
----------+-------------------------------------------------------------
```

```
/tanhdelta |  -.1201056    .0150796                    -.1496611   -.0905502
-----------+----------------------------------------------------------------
     delta |  -.1195314    .0148641                    -.1485536   -.0903035
----------------------------------------------------------------------------
Likelihood-ratio test of delta=0:  chi2(1) = 48.81  Prob>=chi2 = 0.0000
. abich
AIC Statistic  =    3.997356            AIC*n     =  7922.7588
BIC Statistic  =    4.000431            BIC(Stata) =  7950.7183
```

The AIC and BIC statistics are reduced, and the value of the dispersion parameter, delta, is negative, indicating the dispersion parameter has adjusted for underdispersion. Note that /tanhdelta is the inverse hyperbolic tangent function of delta, with a definition of

$$\text{tanhdelta} = \frac{1}{2}\log\left(\frac{1+\delta}{1-\delta}\right) \tag{8.2}$$

The dispersion parameter for the generalized Poisson is

$$\phi = \frac{1}{(1-\delta)^2} \tag{8.3}$$

When $\delta = 0$, the model is equidispersed (i.e., it is Poisson). If $\delta > 0$, the model is overdispersed; if $\delta < 0$, the model is underdispersed. A likelihood ratio test of $\delta = 0$ is similar to the likelihood ratio test of the negative binomial, which is a test of $\alpha = 0$. In this sense, $\alpha = \delta$. p-values less than 0.05 favor the generalized Poisson over the Poisson.

Zero-inflated generalized Poisson (ZIGP) models are appropriate for modeling any type of extradispersed count data. Negative values of delta adjust for Poisson underdispersion; positive values adjust for Poisson overdispersion. I will revert to an evaluation of the **rwm1984** data, which are overdispersed:

```
. zigp docvis outwork age, nolog inflate(outwork age) vuong zip eform

Zero-inflated generalized Poisson regression    Number of obs   =      3874
Regression link:                                 Nonzero obs     =      2263
Inflation link : logit                           Zero obs        =      1611
                                                 LR chi2(2)      =     84.55
Log likelihood = -8272.747                       Prob > chi2     =    0.0000
----------------------------------------------------------------------------
    docvis |   exp(b)    Std. Err.     z    P>|z|    [95% Conf. Interval]
-----------+----------------------------------------------------------------
docvis     |
   outwork |  1.369381   .1089162    3.95   0.000    1.171716    1.600391
```

```
        age |  1.014631   .0032109    4.59   0.000    1.008357    1.020944
      _cons |   1.79139   .2550683    4.09   0.000    1.355162     2.36804
------------+----------------------------------------------------------------
inflate     |
    outwork |  -.3160915   .2510973   -1.26   0.208    -.8082332    .1760502
        age |  -.0194765   .0093451   -2.08   0.037    -.0377926   -.0011604
      _cons |  -.3925934    .398819   -0.98   0.325    -1.174264    .3890774
------------+----------------------------------------------------------------
atanhdelta  |
      _cons |   .7741937   .0171378   45.17   0.000     .7406042    .8077831
------------+----------------------------------------------------------------
      delta |   .6493614   .0099113                     .6295101    .6683655
------------+----------------------------------------------------------------
LR test of delta=0 (zigp versus zip):        X2 =    7785.56   Pr>X2= 0.0000
Vuong test of zigp vs. gen Poisson:           z =       3.89   Pr>z = 0.0000
    Bias-corrected (AIC) Vuong test:          z =       3.50   Pr>z = 0.0002
    Bias-corrected (BIC) Vuong test:          z =       2.28   Pr>z = 0.0114

. abic
AIC Statistic   =    4.274521         AIC*n       = 16559.494
BIC Statistic   =    4.277939         BIC(Stata)  = 16603.328
```

Predicted counts and zeros are obtained by using the **gpois_p** and **predict** commands.

```
. gpois_p prgc, n
. predict prg0, pr
. sum docvis  pr*

    Variable |       Obs        Mean    Std. Dev.       Min        Max
-------------+-----------------------------------------------------------
      docvis |      3874    3.162881    6.275955          0        121
        prcg |      3874    3.930725    1.020594   2.575682   6.214907
        prc0 |      3874    .2074883    .0466571   .1239893    .2932782
```

The likelihood ratio test of $\delta = 0$ at the bottom of the output assesses the scale parameter of the counting process. In effect, this test determines whether there is a significant preference for ZIGP over the ZIP model for these data. ZIGP is preferred.

The Vuong test assesses whether there is a significant preference for the ZIGP model over the generalized Poisson (GP). The test statistic is normally distributed, with large positive values favoring the ZIGP model and large

- Counts are erratically distributed and do not appear to follow a parametric count distribution very well. (quantile count models)
- Data are longitudinal or clustered in nature, where observations are not independent. (panel models; e.g., GEE, GLMM)
- Data are nested in levels or are in a hierarchical structure. (multilevel models)

A very brief overview of the Bayesian modeling of count data will be presented in Section 9.8.

9.1 SMALL AND UNBALANCED DATA – EXACT POISSON REGRESSION

Exact statistics is a highly iterative technique that uses the conditional distributions of the sufficient statistics of the model parameters, assuring that the distribution is completely determined. Many models involve calculation of thousands of permutations. Exact statistics therefore is not a maximum likelihood method, which relies on asymptotic standard errors to determine the significance of predictors. Cytel's LogXact software is the foremost statistical package that provides exact statistics for a wide variety of logistic and Poisson models. If you have access to that package, I recommend it. Stata and SAS also have exact statistics capabilities, as demonstrated herein.

As an example, I have selected a random subset of the 1991 Arizona Medicare data for patients hospitalized subsequent to undergoing a CABG (DRGs 106, 107) or PTCA (DRG 112) cardiovascular procedure.

The data include several possible explanatory predictors, but only two appear to significantly contribute to understanding and predicting the length of hospital stay following surgery. The predictor of foremost interest is *procedure*, which has a value of 1 if the patients had a CABG, and 0 if they had a PTCA. *Procedure* is adjusted by *type*, which is coded as 1 if the patients are admitted as an emergency or urgent patients and 0 if they are admitted on an elective basis, which is a nonemergency classification. Patient *age* ranges from 65 to 79.

Response: *los* – length (in days) of hospital stay following surgery

Predictors: *procedure*: 1 = CABG; 0 = PTCA

type: 1 = emergency/urgent admission; 0 = elective admission

A tabulation of length of stay (*los*) for these data is given as

```
. use azcabgptca
. tab los
        LOS |       Freq.      Percent         Cum.
------------+-----------------------------------
          1 |          11        17.46        17.46
          2 |          23        36.51        53.97
          3 |          18        28.57        82.54
          4 |           4         6.35        88.89
          5 |           4         6.35        95.24
          6 |           3         4.76       100.00
------------+-----------------------------------
      Total |          63       100.00
```

A cross-tabulation of *procedure* and *type* shows that of the 63 records, no emergency/urgent admissions were given a CABG. The fact that patients are generally given a CABG when they have a more serious cardiovascular condition appears to conflict with the data. But since we have data on only three CABG patients, the data are not likely representative of the underlying population of cardiovascular procedures. However, they are the only data we have to work with for this study:

```
. tab procedure type
    1=CABG; |        Severity
    0=PTCA  |  Elective   Emer/Urg |      Total
-----------+----------------------+----------
      PTCA |        34         26 |         60
      CABG |         3          0 |          3
-----------+----------------------+----------
     Total |        37         26 |         63
```

Since the goal of our study is to determine comparative length of stay between the two procedures, we should first tabulate *los* on *procedure* to obtain a descriptive look at the relationship:

```
. tab los procedure
            |    1=CABG; 0=PTCA
        LOS |       PTCA       CABG |      Total
-----------+----------------------+----------
          1 |         11          0 |         11
          2 |         23          0 |         23
```

```
         3 |          17          1 |          18
         4 |           4          0 |           4
         5 |           3          1 |           4
         6 |           2          1 |           3
-----------+----------------------+----------
     Total |          60          3 |          63
```

There are few observations for CABG patients, but it does appear that PTCA patients stay in the hospital postsurgery for less time on average than CABG patients:

```
. sort procedure
. by procedure: sum los
PTCA
      Variable |        Obs        Mean    Std. Dev.        Min        Max
-------------+---------------------------------------------------------
           los |         60    2.516667    1.214205          1          6
CABG
      Variable |        Obs        Mean    Std. Dev.        Min        Max
-------------+---------------------------------------------------------
           los |          3    4.666667    1.527525          3          6
```

The joint mean is 2.62.

```
. sum los
      Variable |        Obs        Mean    Std. Dev.        Min        Max
-------------+---------------------------------------------------------
           los |         63    2.619048    1.300478          1          6
```

Since *los* is a count, the data are modeled using Poisson regression. The model is parameterized so that incidence rate ratios are displayed in place of coefficients:

```
R
================================================================
library(COUNT); library(COUNT)
data(azcabgptca); attach(azcabgpca)
table(los); table(procedure, type); table(los, procedure)
summary(los)
summary(c91a <- glm(los ~ procedure+ type, family=poisson, data=azcabgptca))
modelfit(c91a)
summary(c91b <- glm(los ~ procedure+ type, family=quasipoisson, data=azcabgptca))
```

```
modelfit(c91b)
library(sandwich); sqrt(diag(vcovHC(c91a, type="HC0")))
=================================================================

. glm los procedure type, fam(poi) eform nolog

Generalized linear models                    No. of obs      =        63
Optimization     : ML                        Residual df     =        60
                                             Scale parameter =         1
Deviance       =   30.23003933               (1/df) Deviance =  .503834
Pearson        =   29.12460512               (1/df) Pearson  = .4854101
Variance function: V(u) = u                         [Poisson]
Link function    : g(u) = ln(u)                     [Log]

                                             AIC             =  3.335811
Log likelihood   = -102.0780388              BIC             = -218.358
-------------------------------------------------------------------------
             |                 OIM
         los |       IRR   Std. Err.      z     P>|z|    [95% Conf. Interval]
-------------+-----------------------------------------------------------------
   procedure |  2.144144   .6249071     2.62   0.009     1.21108    3.796078
        type |  1.360707   .2215092     1.89   0.058     .989003    1.872111
       _cons |  2.176471   .2530096     6.69   0.000    1.733016    2.733399
-------------------------------------------------------------------------

. abic
AIC Statistic  =    3.335811        AIC*n      = 210.15608
BIC Statistic  =    3.345202        BIC(Stata) = 216.58548
```

It appears that the data are severely underdispersed, as well as unbalanced. The response variable does not allow zero counts, and its mean value is low; however, this typically results in overdispersion, not underdispersion. The foremost reason for the underdispersion is the fact that some 54% of counts are either 1 or 2; 85% of the values in the data have counts of 3 or less. This clumping effect results in underdispersion. If the remaining count values were skewed farther to the right, the mean would also shift to a higher value, if only by 1 or perhaps less. But this may help eliminate such overt underdispersion in the data.

Given the extradispersion in the data, we should either employ scaling to the standard errors or submit them to adjustment using a robust or sandwich estimator. Scaling by the Pearson dispersion statistic is the same as using the quasipoisson "family" with R's **glm** function. Since the heading statistics are the same as the previous displayed model output, the nohead option is used to exclude them from being printed:

```
. glm los procedure type ,  fam(poi) eform scale(x2) nolog nohead
-----------------------------------------------------------------------------
            |               OIM
       los |      IRR   Std. Err.      z    P>|z|     [95% Conf. Interval]
-----------+-----------------------------------------------------------------
 procedure |  2.144144   .4353814     3.76   0.000     1.440165     3.19224
      type |  1.360707   .1543285     2.72   0.007     1.08949     1.699441
     _cons |  2.176471   .1762753     9.60   0.000     1.857004    2.550896
-----------------------------------------------------------------------------
```

(Standard errors scaled using square root of Pearson X2-based dispersion.)

Adjustment of the standard errors using a robust estimator may be obtained using the following code:

```
. glm los procedure type,  fam(poi) eform vce(robust) nolog nohead
```

The predictor statistics are nearly the same as for the scaled model. This is a common result. Most statisticians tend to prefer using robust adjustment rather than scaling, but the results are typically the same.

We should check the zero-truncated Poisson model results since the data are such that zero counts are excluded and the mean of the response variable is low. The ZTP model may fit the data better than the traditional Poisson model:

```
R
================================================================
library(gamlss.tr)
gen.trun(0,"POI", type="left", name="leftr")
summary(c91c <- gamlss(los~ procedure+type, data=az
ptca, family=POleftr))
================================================================
```

```
. ztp los procedure type ,  nolog irr vce(robust)
-----------------------------------------------------------------------------
            |             Robust
       los |      IRR   Std. Err.      z    P>|z|     [95% Conf. Interval]
-----------+-----------------------------------------------------------------
 procedure |  2.530652   .4607842     5.10   0.000     1.771106    3.615932
      type |  1.521065   .2296154     2.78   0.005     1.131496    2.044761
     _cons |  1.825901   .1480402     7.43   0.000     1.557628     2.14038
-----------------------------------------------------------------------------
. abic
AIC Statistic  =   3.131446          AIC*n     = 197.28108
BIC Statistic  =   3.140837          BIC(Stata) = 203.71049
```

The fit is indeed superior – by 13 points using AIC*n.

Both the traditional and zero-inflated models should be checked for mis-specification. The Poisson model produces the following result, indicating that the model is properly specified – or rather, not likely to be misspecified. The top output tests the Poisson model and the second tests the ZIP.

```
. linktest  /* partial display of output */
-------------------------------------------------------------------------
    los |      Coef.   Std. Err.      z    P>|z|     [95% Conf. Interval]
--------+----------------------------------------------------------------
 _hatsq |   1.574352    2.691572    0.58   0.559    -3.701031    6.849735
--------+----------------------------------------------------------------
 _hatsq |  -3.11e-07    .9064075   -0.00   1.000    -1.776526    1.776526
-------------------------------------------------------------------------
```

The bottom *_hatsq* coefficient approximates zero; the *p*-value is 1. The ZIP model appears to be misspecified. This likely results from the fact that the model excludes zero counts yet is rather severely underdispersed.

Using an exact statistics procedure on these data produces the following coefficients and *p*-values. Remember that the *p*-values and confidence intervals are not based on asymptotic statistics. They are of the same nature as the exact Fisher test when evaluating tables. We see that *procedure* is significant at the 95% level but *type* is not. The standard Poisson model has *p*-values of 0.09 and 0.058 for *procedure* and *type*, respectively. When scaled – and when a robust sandwich estimator is applied – the *p*-values are 0.008 and 0.005, respectively. We usually think when there is extradispersion in the data that scaling (quasipoisson) or applying a robust estimator produces more accurate standard errors, and hence *p*-values/confidence intervals. But here it is clear that they do not. The unadjusted Poisson standard errors and *p*-values are closer to the exact values than are the scaled and sandwich results. The difference is important since the exact values inform us that *type* is not a significant predictor:

```
. expoisson los procedure type, irr
Exact Poisson regression
                                     Number of obs =          63

-------------------------------------------------------------------------
       los |        IRR    Suff.  2*Pr(Suff.)     [95% Conf. Interval]
-----------+-------------------------------------------------------------
 procedure |   2.144144      14       0.0224     1.118098    3.828294
      type |   1.360707      77       0.0701     .9760406    1.898079
-------------------------------------------------------------------------
```

	TABLE 9.1. Cytel LogXact Exact Statistics						
Parameter				Confidence Interval and *p*-Value for Beta			
Estimates	Point Estimate				95	%CI	2*1-sided
Model Term	Type	Beta	SE(Beta)	Type	Lower	Upper	*p*-Value
PROCEDURE	CMLE	0.7627	0.2914	Exact	0.1116	1.342	0.02237 <==
TYPE	CMLE	0.308	0.1628	Exact	−0.02425	0.6408	0.07008 <==

I have run a LogXact model (see Table 9.1) on the same data. Note that the conditional maximum likelihood (exact) estimates are the same for both *procedure* and *type*. Exponentiate the LogXact coefficients and confidence intervals to obtain the same values. For example, $\exp(.7627) = 2.144$ and $\exp(.308) = 1.3607$.

At the time of this writing, no suitable R code exists for exact Poisson regression. Exact Poisson regression is recommended for all models that are unbalanced as well as for data with few observations. Memory is a problem with larger models, but exact statistics software now provides alternative models that generally approximate exact values; these include median unbiased estimates (Stata and SAS) and Monte Carlo estimation (LogXact). As computers become faster, exact methods will likely find more widespread use.

9.2 MODELING TRUNCATED AND CENSORED COUNTS

There are many times when certain data elements are lost, discarded, ignored, or otherwise excluded from analysis. Truncated and censored models have been developed to deal with these types of data. Both models take three forms: truncation or censoring from below, truncation or censoring from above, and truncation or censoring at the endpoints of an interval of counts. Count model forms take their basic logic from truncated and censored continuous response data, in particular from Tobit (Amemiya 1984) and censored normal regression (Goldberger 1983), respectively.

Count sample selection models also deal with data situations in which the distribution is confounded by an external condition. We do not address this type of model in this book. See Hilbe (2011) for details.

Censored and truncated count models are related, with only a relatively minor algorithmic difference between the two. The essential difference relates to how response values beyond user-defined cut points are handled. Truncated models eliminate the values altogether; censored models revalue them to the value of the cut point. In both cases, the probability function and log-likelihood functions must be adjusted to account for the change in the distribution of the response. We begin by considering truncation.

9.2.1 Truncated Count Models

In order to understand the logic of truncation, we begin with the basic Poisson probability mass function, defined earlier as

$$\text{Prob}\,(Y = y) = \frac{e^{-\mu_i}\mu_i^{y_i}}{y_i!}, \quad y = 0, 1, \dots$$

When we discussed zero-truncated Poisson models in Section 7.1, we adjusted the preceding Poisson distribution to account for the structural absence of zeros. We discovered that the probability of a zero count for the Poisson distribution is $\exp(-\mu)$, for the negative binomial $(1-\alpha\mu)^{-1/2}$, and for the PIG $\exp((1/\alpha)*(1-\text{sqrt}(1+2/(\alpha\mu))))$. Other zero-truncated models were addressed, each based on dividing the underlying model PDF by the respective probability of 0. I will focus on using truncated Poisson models, but understand that the same logic applies to other distributions.

When truncating a zero count from the Poisson distribution, the probability of zero is subtracted from 1, and the result is divided into the full Poisson PDF; that is, $\text{PDF}/(1-\exp(-\mu))$. The same is the case for other distributions, except that the formula differs.

Going farther from 0, the Poisson probability of 1 is $\mu*\exp(-\mu)$. If truncation is at 1, then both 0 and 1 must be excluded from the distribution and an adjustment for both must be made in the resulting adjusted PDF. We do this by summing the two probabilities and subtracting the sum from 1. This value is divided into the Poisson PDF:

$$\text{Prob}\,(Y = (y = 0, 1)) = \frac{e^{-\mu_i}\mu_i^{y_i}}{(1 - (e^{-\mu_i} + \mu_i e^{-\mu_i}))\, y_i!}, \quad y = 2, 3, \dots \quad (9.1)$$

This distribution can be called a left-truncated at 1 Poisson distribution. When establishing a left truncation at point 1, we place a cut point, C, at 1, and the first number to be in the nontruncated distribution is $C + 1$.

The same logic applies to each greater integer or count from the left. Left refers to the number line beginning with 0 and moving one integer to the right with each higher count:

```
----------------------------------------------------  -------------->>
  ^       ^       ^       ^       ^      ^      ^       ^    ...    ^       ^
  0       1       2       3       4      5      6       7        n-1       n
```

Example: Left-truncated at 3: model includes counts from 4 to the highest number in the model.

Right-truncated at 10: model includes counts from 0 to 9.

The left-truncated Poisson PDF in general is Poisson PDF/Prob$(y > C)$. Numerically, this appears as

$$\text{Prob}\,(Y = y | Y > C) = \frac{\dfrac{\exp{(-)}\,\mu^{y}}{y!}}{1 - \sum_{j=0}^{C} \dfrac{\exp{(-\mu)}\,\mu^{j}}{j!}}, \quad \text{for } y = C+1, C+2, \ldots$$

$$(9.2)$$

When a cut point is on the right side, it is the higher-value end of the sorted distribution:

$$\text{Prob}\,(Y = y | Y < C) = \frac{e^{-\mu_i}\,\mu_i^{y_i}/y_i!}{\text{Prob}\,(y_i < C)} = \frac{e^{-\mu_i}\,\mu_i^{y_i}/y_i!}{\sum_{j=0}^{C-1} e^{-\mu_i}\,\mu_i^{j_i}/j_i!},$$

$$\text{for } y = 0, 1, \ldots, C-1 \qquad (9.3)$$

An example will make it clear how to actually use the model. The same German health reform data set, **rwm1984**, is used for our example data. Employing the user-created Stata **treg** command (Hardin and Hilbe 2014b), I first display the results of estimating a left-truncated at 3 Poisson model (see R code in Table 9.2). With a left cut point at 3, counts in the resultant model begin at 4, or $C + 1$ (see *distribution* in header). Limdep and the **gamlss** package in R are the only other software programs I know of that provide the capability of modeling truncated count data, although they are both limited to the Poisson and negative binomial distributions. Counts 3 and above are being modeled. Note that **treg** allows the user to model left-, right-, and interval-truncated Poisson, negative binomial, generalized Poisson, PIG, NB-P, and three-parameter generalized Famoye negative binomial regressions. The latter four models are new additions with the **treg** command. The truncated PIG

TABLE 9.2. R: Left-Truncated at 3 Poisson

```
library(COUNT); data(rwm5yr); rwm1984 <- subset(rwm5yr, year==1984)
summary(plt <- gamlss(docvis~outwork + age,data=rwm1984,family=PO))
library(gamlss); library(gamlss.tr); pltvis<-subset(rwm1984, rwm1984$docvis>3)
summary(ltpo <- gamlss(docvis~outwork+age, family=trun(3, "PO", "left"),
data=pltvis))
----------------
pltvis<-subset(rwm1984, rwm1984$docvis>3)       # alternative method
gen.trun(3, "PO", "left")                        # saved globally for session
summary(lt3po <- gamlss(docvis~outwork+age, family=POleft, data=pltvis))
```

is actually the best-fitted model of the alternatives. I suggest using robust or sandwich standard errors with truncated and censored models. In the following example, model SEs indicate that *outwork* is significant when it is not.

```
. treg docvis outwork age if docvis>3, dist(poisson) ltrunc(3) nolog vce(robust)

Truncated Poisson regression              Number of obs   =        1022
Distribution = {4, ..., .}                LR chi2(2)      =       69.68
Log pseudolikelihood =  -4802.67          Prob > chi2     =      0.0000
-----------------------------------------------------------------------------
             |               Robust
      docvis |      Coef.   Std. Err.      z    P>|z|     [95% Conf. Interval]
-------------+---------------------------------------------------------------
     outwork |   .0735456   .0733219     1.00   0.316    -.0701626    .2172539
         age |    .006175   .0029573     2.09   0.037     .0003789    .0119712
       _cons |   1.923062    .129157    14.89   0.000     1.669919    2.176205
-----------------------------------------------------------------------------
```

For a truncation from the right side (i.e., at the higher end of the distribution), I display an example of a right-truncated at 10 PIG. With right truncation at 10, only counts ranging from 0 to 9 are included in the model. Note the distribution shown in the header information:

```
. treg docvis outwork age if docvis<10, dist(pig)  rtrunc(10)  nolog vce(robust)

Truncated Poisson IG regression           Number of obs   =        3566
Distribution = {0, ..., 9}                LR chi2(2)      =      176.44
Log pseudolikelihood = -6528.991          Prob > chi2     =      0.0000
```

```
-------------------------------------------------------------------------------
             |               Robust
      docvis |      Coef.   Std. Err.      z    P>|z|     [95% Conf. Interval]
-------------+-----------------------------------------------------------------
     outwork |    .415145   .0506344     8.20   0.000     .3159033    .5143867
         age |    .018404   .0022195     8.29   0.000     .0140538    .0227542
       _cons |  -.3912701   .1007169    -3.88   0.000    -.5886716   -.1938686
-------------+-----------------------------------------------------------------
    /lnalpha |    .5445644   .052255                      .4421464    .6469824
-------------+-----------------------------------------------------------------
       alpha |   1.723857   .0900802                     1.556044    1.909769
-------------------------------------------------------------------------------
```

The example in Table 9.3 provides a right truncation at 10 using the **gamlss** function: It should also be noted that like using zero-truncation when the data have no zero counts, if the data have only low value counts, e.g., 0–6, a better fitted model may be produced by employing a right truncated model to adjust for not having the higher counts assumed by the distribution.

TABLE 9.3. R: Right-Truncated Poisson: Cut = 10

```
rtp<-subset(rwm1984, rwm1984$docvis<10)
summary(rtpo <- gamlss(docvis~outwork + age, data=rtp,
    family=trun(9, "PO", type="right")))
```

Finally, suppose we wish to model visits to the doctor, but only for patients who have in fact visited a physician during the calendar year 1984. Moreover, for our example, suppose also that visits were not recorded for more than 18 visits. We then model the data with a left truncation at 0 and right truncation at 19. This is called *interval truncation*. This time a truncated NB-P model is used to model the interval data, parameterized so that estimates are in terms of incidence rate ratios. This can be achieved with the following Stata code, but it is not yet possible using R or SAS:

```
. treg docvis outwork age if docvis<0 & docvis<19, dist(nbp) ltrunc(0)
rtrunc(19) vce(robust) eform
```

```
Truncated neg. bin(P) regression          Number of obs   =      2172
Distribution = {1, ..., 18}               LR chi2(2)      =     52.70
Log pseudolikelihood = -5007.664          Prob > chi2     =    0.0000
```

```
           |               Robust
  docvis |    exp(b)    Std. Err.      z    P>|z|      [95% Conf. Interval]
---------+----------------------------------------------------------------
 outwork | 1.330828     .0819098     4.64   0.000      1.179593    1.501453
     age | 1.010227     .0027296     3.77   0.000      1.004891    1.015591
   _cons | 1.840426     .2360925     4.76   0.000      1.431281    2.366528
---------+----------------------------------------------------------------
      /P | .539376      .2738796     1.97   0.049      .0025817    1.07617
/lnalpha | 1.593174     .3356481                       .9353162    2.251033
---------+----------------------------------------------------------------
   alpha | 4.91934      1.651167                        2.548019   9.497538
--------------------------------------------------------------------------
```

The predictors appear to contribute significantly to understanding *docvis*. The dispersion parameter is high at 4.92, and the NB-P scale parameter is a significant 0.539. The variance function is then $\mu + 4.92\mu^{.54}$.

9.2.2 Censored Count Models

The censoring that is understood when dealing with count models differs from the censoring that occurs in survival models. A survival-type parameterization for censoring was developed by Hilbe (1998) and is discussed in detail in Hilbe (2011) but will not be addressed here. We will focus on the traditional parameterization of censoring for count models, which is nearly the same as truncation. The difference is that instead of truncated values being excluded from the truncated distribution, censored values are revalued to the value of the cut point. The distinction is subtle but important:

Left censoring: Left: $P(Y \leq C)$,

If $C = 3$, 3 is the smallest value in the model. Values that may have been lower are revalued to C. Any response in the data that is less than 3 is also considered to be less than or equal to 3.

Right censoring: Right: $P(Y \geq C)$,

If $C = 15$, 15 is the highest observed value in the model. Values that may have been greater in the data are revalued to the value at C. Any response in the data that is greater than 15 is also considered to be greater than or equal to 15.

The **cpoissone** (Hilbe 2007a) command is used to model a left-censored at 3 Poisson model. The software is on the book's web site:

```
. gen cenvar=1
. replace cenvar if docvis<=3
```

```
. cpoissone docvis outwork age, censor(cenvar) cleft(3) nolog

Censored Poisson Regression                    Number of obs   =       3874
                                               Wald chi2(2)    =     868.53
Log likelihood = -9698.6316                    Prob > chi2     =     0.0000
------------------------------------------------------------------------------
    docvis |      Coef.   Std. Err.      z    P<|z|     [95% Conf. Interval]
-----------+------------------------------------------------------------------
   outwork |   .2824314   .0176317    16.02   0.000     .247874    .3169888
       age |   .0153682   .0007858    19.56   0.000     .0138281   .0169084
     _cons |   .6101677   .0367009    16.63   0.000     .5382354   .6821001
------------------------------------------------------------------------------
AIC Statistic =      5.009
```

Right censoring is a more common application when using censored Poisson or negative binomial regressions, whereas left truncation is more commonly used with truncation models. **cpoissone** is used here to model the right-censored at 10 model:

```
. replace cenvar=1
. replace cenvar=-1 if docvis>=10
. cpoissone docvis outwork age, censor(cenvar) cright(10) nolog

Censored Poisson Regression                    Number of obs   =       3874
                                               Wald chi2(2)    =     864.81
Log likelihood = -10463.973                    Prob > chi2     =     0.0000
------------------------------------------------------------------------------
    docvis |      Coef.   Std. Err.      z    P>|z|     [95% Conf. Interval]
-----------+------------------------------------------------------------------
   outwork |   .3530813   .0213676    16.52   0.000     .3112017   .3949609
       age |   .0178009   .0009444    18.85   0.000     .0159499   .0196519
     _cons |   -.045936    .04368     -1.05   0.293     -.1315473  .0396752
------------------------------------------------------------------------------
AIC Statistic =   5.404
```

The most common use of a truncation count model is left truncation, in particular left truncation at 0. Censored models are usually applied to some right-censored data. Censored models may also be used in hurdle models. I demonstrated a PIG-Poisson hurdle model earlier in the book, where the Poisson component was created as a right-censored at 1 model. It is possible to have a right-censored at 2 Poisson or negative binomial and a logit binary where counts begin at greater than 2. That is, we created a visit variable at "gen visit = *docvis* > 0" where counts 1 and greater are included. We could, however, have set *docvis* >1 so that the binary component would include

TABLE 9.4. R: Left-Censored Poisson at Cut = 3

```
library(gamlss.cens); library(survival); library(COUNT)
data(rwm5yr); rwm1984 <- subset(rwm5yr, year==1984)
lcvis <- rwm1984
cy <- with(lcvis, ifelse(docvist<3, 3, docvis))
ci <- with(lcvis, ifelse(docvis<=3, 0, 1))
Surv(cy,ci, type="left")[1:100]
cbind(Surv(cy,ci, type="left")[1:50], rwm1984$docvis[1:50])
lcmdvis <- data.frame(lcvis, cy, ci )
rm(cy,ci); gen.cens("PO",type="left")
lcat30<-gamlss(Surv(cy, ci, type="left") ~ outwork + age,
     data=lcvis, family=POlc)
summary(lcat30)
```

0 and 1 combined, for a right-censored at 2 count model. This would give us a hurdle model with an extended binary component. This discussion brings us to finite mixture models.

9.2.3 Poisson-Logit Hurdle at 3 Model

An extended type of hurdle model that branches over to finite mixture models, which we discuss next, can be created by setting the hurdle at a higher place in the range of counts than at 0. Here I show an example of a Poisson-logit hurdle that is set at a count of 3. To execute this, I create a right-censored at 3 Poisson model and a logit model that has a response with 1's for counts 3 and above and 0's for counts 0, 1, and 2. See the tabulation of *docvis* and a variable called *cenvar* that follows. The binary and count components constitute the Poisson-logit hurdle model with the hurdle point at 3, not at 0 as usual. This logic can be applied to the other distributions as well to give a wide range of distributions and cut points reflecting the data being modeled.

TABLE 9.5. R: Right-Censored Poisson at 10

```
library(gamlss.cens); library(survival); rcvis <- rwm1984
cy <- with(rcvis, ifelse(docvis>=10, 9, docvis))
ci <- with(rcvis, ifelse(docvis>=10, 0, 1))
rcvis <- data.frame(rcvis, cy, ci )
rm(cy,ci) ; gen.cens("PO",type="right")
summary(rcat30<-gamlss(Surv(cy, ci) ~ outwork + age,
     data=rcmdvis, family=POrc, n.cyc=100))
```

```
POISSON COUNT COMPONENT
. gen cenvar=1
. replace cenvar=-1 if docvis>=3
. epoissone docvis outwork, nolog cright(3)
. cpoissone docvis outwork age, censor(cenvar) cright(3) nolog
```

Censored Poisson Regression   Number of obs   =   3874

                                              Wald chi2(2)   =   393.14

Log likelihood = -5587.6022   Prob > chi2   =   0.0000

----
   docvis | Coef.   Std. Err.   z   P>|z| [95% Conf. Interval]
---------+--
  outwork | .3250068   .0287805   11.29   0.000   .2685981   .3814156
      age | .0164991   .0012707   12.98   0.000   .0140087   .0189896
    _cons | -.4033441   .0580651   -6.95   0.000   -.5171495   -.2895387

AIC Statistic =   2.886

```
LOGIT BINARY COMPONENT
. gen visit = docvis>=3
. tab visit
```

    visit | Freq.   Percent   Cum.
---------+-----------------------------------
        0 | 2,499   64.51   64.51
        1 | 1,375   35.49   100.00
---------+-----------------------------------
    Total | 3,874   100.00

```
. glm visit outwork age, fam(bin) nolog nohead
```

          | OIM
    visit | Coef.   Std. Err.   z   P>|z| [95% Conf. Interval]
---------+--
  outwork | .5701073   .0717644   7.94   0.000   .4294516   .710763
      age | .0287807   .0031698   9.08   0.000   .022568   .0349933
    _cons | -2.10176   .1438936   -14.61   0.000   -2.383786   -1.819733

9.3 COUNTS WITH MULTIPLE COMPONENTS – FINITE MIXTURE MODELS

What do we do if we suspect that the response variable of our model consists of counts that have been generated from different data-generating mechanisms? That is, there are times when the data to be modeled are generated from more than one source. Finite mixture models have been developed to model this sort of data situation (see Table 9.6).

TABLE 9.6. R: Poisson–Poisson Finite Mixture Model

```
library(COUNT)
library(flexmix)
data(fishing)
attach(fishing)
fmm_pg <- flexmix(totabund~meandepth + offset(log(sweptarea)),
data=rwm1984, k=2,
        model=list(FLXMRglm(totabund~., family="NB1"),
                   FLXMRglm(tpdocvis~., family="NB1")))
parameters(fmm_pg, component=1, model=1)
parameters(fmm_pg, component=2, model=1)
summary(fmm_pg)
```

As an example, we'll start with an NB2-NB2 mixture. This means that both components in the data are distributed as NB2, but with differing parameters. We will use the **fishing** (adapted from Zuur, Hilbe, and Ieno 2013) data to determine whether the data appear to be generated from more than one generating mechanism. The data are adapted from Bailey et al. (2008), who were interested in how certain deep-sea fish populations were impacted when commercial fishing began in locations with deeper water than in previous years. Given that there are 147 sites that were researched, the model is (1) of the total number of fish counted per site (*totabund*); (2) on the mean water depth per site (*meandepth*); (3) adjusted by the area of the site (*sweptarea*); and (4) the log of which is the model offset.

```
. use fishing
. fmm totabund meandepth, exposure(sweptarea) components(2) mixtureof(negbin2)

2 component Negative Binomial-2 regression        Number of obs    =        147
                                                  Wald chi2(2)     =     167.04
Log likelihood = -878.91748                       Prob > chi2      =     0.0000
------------------------------------------------------------------------------
    totabund |      Coef.   Std. Err.      z    P>|z|     [95% Conf. Interval]
-------------+----------------------------------------------------------------
component1   |
   meandepth |  -.0008543   .0001498    -5.70   0.000    -.0011479   -.0005606
       _cons |  -4.144922   .5319592    -7.79   0.000    -5.187543   -3.102301
ln(swepta~a) |          1  (exposure)
-------------+----------------------------------------------------------------
component2   |
   meandepth |  -.0011734   .0000919   -12.77   0.000    -.0013535   -.0009932
```

```
       _cons |   -2.816184    .3340012     -8.43   0.000    -3.470815    -2.161554
ln(swepta~a) |           1   (exposure)
-------------+----------------------------------------------------------------
  /imlogitpi1 |     .06434    1.077807      0.06   0.952    -2.048124     2.176804
    /lnalpha1 |   -.4401984    .2589449     -1.70   0.089     -.9477211     .0673244
    /lnalpha2 |   -1.488282    .4407883     -3.38   0.001     -2.352212     -.624353
-------------+----------------------------------------------------------------
      alpha1 |    .6439087    .1667369                        .3876234     1.069642
      alpha2 |    .2257601    .0995124                        .0951585     .5356078
         pi1 |    .5160794    .2691732                        .1142421     .898147
         pi2 |    .4839206    .2691732                         .101853     .8857579
-------------------------------------------------------------------------------
```

The probability of a site being in the first component of the distribution is .516, whereas the probability of being in the second component is .484. The confidence intervals do not include 0, so we may consider that both significantly represent the underlying distributional probabilities. It is clear that there is more variability in the first component:

```
. predict mean1, equation(component1)
. predict mean2, equation(component2)
. sum mean*
    Variable |        Obs        Mean    Std. Dev.        Min         Max
-------------+--------------------------------------------------------
   meandepth |        147    2413.088    1254.587        804        4865
       mean1 |        147    150.9075    94.06647    19.26319    415.0464
       mean2 |        147    335.3274    269.7885    15.90274    1128.203
```

The second component has a mean value of fish per site that is some two-and-a-quarter times (335/151) greater than the first component, even though it is only 48% of the overall population. The overall mean fish per site can be calculated as

```
. di .5160794 * 150.9075 + .4839206 * 335.3274
240.15209
```

If the data are partitioned into three components, the probability values for being in each component are

```
         pi1 |    .2230491    .2346794                        .0197984     .8031637
         pi2 |    .7169981    .2227915                        .2275204     .9561276
         pi3 |    .0599529    .0351927                       -.0090236     .1289294
```

This time, the second component has 72% of the data, the first has 22%, and the third 6%. The third component, however, is not significant (confidence interval includes 0). Greater partitions also fail to show more than two significant components. Interestingly, the model used is pooled over precommercial fishing (1977–1989) and commercial fishing (2000–2002) years, although there is some overlap in *period* and *meandepth*:

```
. tab period
0=1977-89;1 |
    =2000+ |      Freq.      Percent        Cum.
------------+-----------------------------------
 1977-1989 |         97        65.99       65.99
 2000-2002 |         50        34.01      100.00
------------+-----------------------------------
     Total |        147       100.00
```

Modeling the data using negative binomial regression but including the binary predictor, *period*, produces a well-fitted model. Using robust estimators yields standard errors that are nearly identical to the following standard model:

```
. nbreg totabund meandepth period, exposure(sweptarea) nolog
----------------------------------------------------------------------------
    totabund |    Coef.   Std. Err.      z    P>|z|     [95% Conf. Interval]
-------------+--------------------------------------------------------------
   meandepth | -.0010168   .0000505   -20.14   0.000    -.0011158   -.0009179
      period | -.4269806   .1260863    -3.39   0.001    -.6741052   -.1798559
       _cons | -3.307003   .1395975   -23.69   0.000    -3.580609   -3.033397
ln(swepta~a) |        1   (exposure)
-------------+--------------------------------------------------------------
     /lnalpha | -.6697869   .1118123                    -.8889351   -.4506388
-------------+--------------------------------------------------------------
       alpha |  .5118176   .0572275                     .4110933    .637221
----------------------------------------------------------------------------
Likelihood-ratio test of alpha=0:  chibar2(01) = 1.4e+04 Prob>=chibar2 = 0.000
```

Finite mixture models may have multiple components and be structured such that the response is comprised of more than one distribution.

9.4 ADDING SMOOTHING TERMS TO A MODEL – GAM

Generalized additive models (GAMs) are a class of models based on generalized linear models for which the linear form of the model, $\Sigma\ x\beta$, is replaced

TABLE 9.7. R: GAM

```
library(COUNT); library(mgcv)
data(rwm5yr); rwm1984 <- subset(rwm5yr, year==1984)
summary(pglm <- glm(docvis ~ outwork + age + female + married +
    edlevel2 + edlevel3 + edlevel4, family=poisson, data=rwm1984))
summary(pgam <- gam(docvis ~ outwork + s(age) + female + married +
    edlevel2 + edlevel3 + edlevel4, family=poisson, data=rwm1984))
plot(pgam)
```

by a sum of smoothed functions, $\Sigma\ s(X)$. The method is used to discover non-linear covariate effects that may not be detectable using traditional statistical techniques.

GAM methodology was originally developed by Stone (1985) but was later popularized by Hastie and Tibshirani (1986, 1990). The authors even developed a software package called GAIM, for generalized additive interactive modeling, which stimulated its inclusion in several of the major commercial packages.

The key concept in GAM modeling is that the partial residuals of continuous predictors in a model are smoothed using a cubic spline, loess smoother, or another type of smoother while being adjusted by the other predictors in the model. The parameters of the smooths are related to the bandwidth that was used for the particular smooth. The relationship that is traditionally given for the GAM distribution is

$$y = \beta_0 + \sum_{i=1}^{j} f_j\left(X_j\right) + \varepsilon \qquad (9.4)$$

It is important to remember that the purpose of using GAM is to determine the appropriate transformation needed by a continuous predictor in order to affect linearity. A GAM employs the *iteratively reweighted least squares* (IRLS) algorithm used in GLM models for estimation. At each iteration, the partial residuals of each relevant continuous predictor in the model are smoothed. Partial residuals are used since they remove the effect of the other model predictors.

I will use as an example the **rwm1984** data, showing the parameter estimates of a GLM and GAM on the full model data. *age* is the only continuous predictor, so the smooth is of it. (See Table 9.7.)

Most models give results similar to those of negative binomial and PIG. R does not support the quantile count model at this time.

9.6 A WORD ABOUT LONGITUDINAL AND CLUSTERED COUNT MODELS

Count models are standard components of the major longitudinal models, as well as multilevel models. There are two broad types of models that are commonly used for longitudinal and multilevel modeling. The first is generalized estimating equations (GEEs), which is a population-averaging method of estimation. GEE models are not true likelihood-based models but rather are examples of quasi-likelihood models. The other type of model is referred to as a subject-specific model. Most random- and mixed-effects models are in this class of models.

9.6.1 Generalized Estimating Equations (GEEs)

GEE models are an extension of the generalized linear model (GLM) where the variance function is adjusted using a correlation matrix. Several standard correlation structures are used for GEE analysis:

- independence – an identity matrix, no correlation effect is specified at all;
- exchangeable – a common correlation value provided to each panel or cluster in the data;
- autoregressive – lag correlation for longitudinal and other time-dependent variables;
- unstructured – separate correlation values for each panel or cluster in the data.

Other structures also exist (stationary, nonstationary, Markov, family) but are rarely used in research. Stata's **xtgee** command is a full GEE package. R has several GEE functions: **geepack**, **glmgee**, and **yags**. The first two are used most often for GEE modeling among R users. The **Genmod** procedure in SAS is a GLM procedure but has an option for modeling GEEs. I can recommend Hardin and Hilbe (2013b), the standard text on the subject. It has both Stata and R examples throughout the book, as well as supplementary SAS code.

TABLE 9.8. R: GEE

```
library(COUNT); library(gee); data(medpar)
summary(pgee <- gee(los ~ hmo + white + age80 + type2 + type3,
                data=medpar,    id=medpar$provnum,
                corstr='exchangeable', family=poisson))
```

To give a brief example, we'll use the U.S. Medicare data called **medpar**. Length of stay is the response, with *hmo*, *white*, *age80*, and types of admission as predictors. Stata requires that the cluster variable be numeric. The **encode** command converts a string to numeric. An exchangeable correlation structure was selected since we do not know anything about the data other than that they are overdispersed and come in panels (i.e., patients within hospitals). The idea is that patients are likely to be treated more alike within hospitals than between hospitals, adding overdispersion to the data. A Poisson GEE model is used for the example. (See Table 9.8.)

```
. use medpar
. encode provnum, gen(hospital)
. xtgee los hmo white age80 type2 type3, i(hospital) c(exch) vce(robust)
  fam(poi) eform
```

```
GEE population-averaged model         Number of obs      =      1495
Group variable: hospital              Number of groups   =        54
Link: log                             Obs per group: min =         1
Family: Poisson                       avg                =      27.7
Correlation: exchangeable             max                =        92
                                      Wald chi2(5)       =     27.51
Scale parameter: 1                    Prob > chi2        =    0.0000
(Std. Err. adjusted for clustering on hospital)
```

los	Coef.	Robust Std. Err.	z	P>|z|	[95% Conf. Interval]	
hmo	-.0876095	.0584452	-1.50	0.134	-.20216	.026941
white	-.1078697	.0732792	-1.47	0.141	-.2514942	.0357549
age80	-.0583864	.0611035	-0.96	0.339	-.1781471	.0613743
type2	.2295517	.0602295	3.81	0.000	.111504	.3475993
type3	.5402621	.1974048	2.74	0.006	.1533559	.9271684
_cons	2.310611	.074003	31.22	0.000	2.165568	2.455654

```
Coefficients:
               Estimate Naive S.E.   Naive z Robust S.E.   Robust z
(Intercept)  2.31069563 0.07615419 30.342331  0.07328592 31.5298719
hmo         -0.08757282 0.06332121 -1.382993  0.05788583 -1.5128541
white       -0.10802330 0.07240366 -1.491959  0.07260164 -1.4878908
age80       -0.05837643 0.05177317 -1.127542  0.06053692 -0.9643112
type2        0.22951223 0.05663849  4.052231  0.05967494  3.8460402
type3        0.54096501 0.08222260  6.579274  0.19567264  2.7646431

Estimated Scale Parameter:  6.504578

The incidence rate ratios are:

> exp(pgee$coefficient)
(Intercept)         hmo       white       age80       type2       type3
 10.0814351   0.9161522   0.8976067   0.9432948   1.2579863   1.7176636
```

9.6.2 Mixed-Effects and Multilevel Models

Mixed-effects models are panel models that are combinations of fixed and random effects, both of which have models named for them. *Fixed effects* are generally regarded as emphasizing the measurements of the effects themselves. That is, what is being measured is important in itself; it is fixed. Standard regression coefficients are fixed effects in that sense. This is distinct from random effects, which assume the measurements are samples or representatives of a greater population. *Random effects* are generally structured so that they are normally distributed with a mean of 0 and standard deviation of σ^2 or $N(0, \sigma^2)$. The random effects themselves are not generally estimated directly but are summarized based on their underlying variance–covariance matrices.

Random-effects models are often divided into two categories – random intercept, and random slopes or coefficients. A random-intercept model is the simplest random-effects model, being structured so that only the intercepts are random. They vary in value across panel intercepts. If the coefficients themselves vary between panels, the model is a random-slopes model.

There are in fact a host of different classifications of fixed- and random-effects models, but these can generally be placed into one of the preceding general categories. We do not have space here to delve into all the varieties

TABLE 9.9. R: Random Intercept Poisson

```
> library(gamlss.mx)
> summary(rip <- gamlssNP(los ~ hmo + white + type2 + type3,
                    random=~1|provnum, data=medpar,
                    family=PO, mixture="gq", K=20))
```

of the models, but we will look at setting up a simple random-intercept
Poisson model using the same data we used for the GEE models:

```
. xtmepoisson los hmo white type2 type3 || provnum:
Mixed-effects Poisson regression          Number of obs      =      1495
Group variable: provnum                   Number of groups   =        54
                                          Obs per group: min =         1
                                                         avg =      27.7
                                                         max =        92
Integration points =    7                 Wald chi2(4)       =    109.81
Log likelihood = -6528.6768               Prob > chi2        =    0.0000
------------------------------------------------------------------------
      los |      Coef.   Std. Err.      z    P>|z|     [95% Conf. Interval]
----------+-------------------------------------------------------------
      hmo | -.0907326   .0259212    -3.50   0.000    -.1415373   -.0399279
    white | -.0275537   .0302745    -0.91   0.363    -.0868906    .0317832
    type2 |  .2319144   .0247704     9.36   0.000     .1833654    .2804634
    type3 |  .1226025   .0488912     2.51   0.012     .0267775    .2184275
    _cons |  2.191951   .0649116    33.77   0.000     2.064726    2.319175
------------------------------------------------------------------------

------------------------------------------------------------------------
  Random-effects Parameters |   Estimate  Std. Err.    [95% Conf. Interval]
----------------------------+-------------------------------------------
provnum: Identity           |
                 sd(_cons)  |  .4116571   .0457474     .3310866    .5118346
------------------------------------------------------------------------
LR test vs. Poisson regression: chibar2(01) =   800.46 Prob>=chibar2 = 0.0000
```

The actual random-intercept values can be obtained using the following
code:

```
. predict raneff, ref
. sum raneff
    Variable |       Obs        Mean    Std. Dev.       Min        Max
-------------+---------------------------------------------------------
      raneff |      1495    .0383374    .2614734   -.7469211    1.37513
```

The R code for the same model is given in Table 9.9:

```
Mu Coefficients:
             Estimate  Std. Error  t value    Pr(>|t|)
(Intercept)   2.11707     0.02839   74.563    0.000e+00
hmo          -0.07501     0.02398   -3.128    1.763e-03
white        -0.03080     0.02774   -1.110    2.670e-01
type2         0.21709     0.02113   10.275    9.996e-25
type3         0.26166     0.03098    8.447    3.115e-17
z             0.50154     0.01642   30.549    7.335e-202
-------------------------------------------------------------------
No. of observations in the fit:  29900
Degrees of Freedom for the fit:  6
      Residual Deg. of Freedom:  1489

Global Deviance:       13086.16
           AIC:        13098.16
           SBC:        13130.02
```

There is a slight difference in estimates between the Stata and R models based on how the models were estimated. Stata is using quadrature, whereas R's **gamlss** is using the EM algorithm. R users often use the **lme4** package for estimation of longitudinal and mixed models. The package includes the longitudinal form of equation (4.16) that was discussed in Section 4.3.1, using the **aictab** function.

Multilevel models are based on the same logic, with levels of nested models being estimated.

You are referred to the following texts for added explanation and examples: Zuur, Hilbe, and Ieno (2013), Hilbe (2011), and Faraway (2006).

A multilevel model is just what you would imagine. Levels are an intrinsic aspect of many data situations. Using the full five-year German health data, we can nest individual patient records within the years from 1984 through 1988. Patients have records from one to all five years. The model used is a three-level random-intercept GLM-Poisson model, which is new to Stata 13 (June 2013):

```
. tab year
year:1984- |
    1988 |       Freq.      Percent        Cum.
------------+-----------------------------------
    1984 |       3,874        19.76       19.76
    1985 |       3,794        19.35       39.10
```

```
         1986 |      3,792       19.34        58.44
         1987 |      3,666       18.70        77.14
         1988 |      4,483       22.86       100.00
    ------------+----------------------------------
        Total |     19,609      100.00
```

```
. meglm docvis outwork age female married   || id: || year:, fam(poi)
Mixed-effects GLM                 Number of obs      =      19609
Family: Poisson
Link:   log
```

```
---------------------------------------------------------------
               |  No. of         Observations per Group
Group Variable |  Groups    Minimum    Average    Maximum
---------------+-----------------------------------------------
           id |    6127          1         3.2          5
         year |   19609          1         1.0          1
---------------------------------------------------------------
```

```
Integration method: mvaghermite              Integration points =      7
                                             Wald chi2(4)       = 585.20
Log likelihood = -41268.492                  Prob > chi2        = 0.0000
```

```
---------------------------------------------------------------------
    docvis |    Coef.    Std. Err.      z    P>|z|   [95% Conf. Interval]
-----------+---------------------------------------------------------------
   outwork |   .1522959   .0318875    4.78   0.000    .0897975    .2147943
       age |    .025162   .0014824   16.97   0.000    .0222565    .0280674
    female |   .4215679     .03611   11.67   0.000    .3507936    .4923422
   married |   -.048628   .0367921   -1.32   0.186   -.1207391    .0234831
     _cons |  -1.108713   .0705329  -15.72   0.000   -1.246955   -.9704716
-----------+---------------------------------------------------------------
id         |
var(_cons) |   1.11585    .035832                    1.047785    1.188337
-----------+---------------------------------------------------------------
id>year    |
var(_cons) |   .8704313   .0202643                    .8316065    .9110686
---------------------------------------------------------------------
LR test vs. Poisson regression: chi2(2) = 70423.53   Prob>chi2 = 0.0000
Note: LR test is conservative and provided only for reference.
```

Id var(_cons) indicates the subject-specific random effects. It is significant with a standard error of .036. There also appears to be more unobserved variability between years than between individuals within each year. The likelihood ratio test informs us that the multilevel model is better fitted than a pooled Poisson model.

9.7 THREE-PARAMETER COUNT MODELS

We have previously examined a three-parameter count model, NB-P, where the power of the second term of the negative binomial variance function is parameterized: $\mu+\alpha\mu^\rho$. When ρ (rho) equals 1, the model is NB1. When $\rho = 2$, the model is NB2, or the traditional version of the negative binomial model. NB-P can be used to select later modeling by NB1 or NB2 regression, or used in its own right as a more malleable means of adjusting for extra correlation in the data. I prefer to use it in the latter manner. However, there is a real sense in which the NB-P is not a true generalized negative binomial model.

A generalized model is typically one that adds another parameter to adjust for correlation or variability in a lower-level model. The negative binomial model is a generalization of the Poisson as a Poisson-gamma mixture. The generalized Poisson model is a type of Poisson–log-normal mixture, and we have discussed the Poisson–inverse Gaussian mixture model (PIG). Each of these models provides a dispersion parameter to adjust for Poisson overdispersion. The generalized Poisson model can also adjust for under-dispersion.

The dispersion parameter for each of these models is singular across counts and observations. But what if the data are such that the dispersion or correlation varies in different parts of the distribution? Another parameter can be provided to adjust for this situation. There are a class of three-parameter models that have been developed to deal with variation in the dispersion statistic by adding a second scale parameter to the distribution. The foremost of these, with developers of note and pertinent references, are

- generalized Waring negative binomial regression (GWRM) (Xekalaki 1983, Rodríguez-Avi et al. 2009)
- generalized Famoye negative binomial regression (NBF) (Famoye 1993)
- extended Poisson process models (EPPMs) (Faddy and Smith 2012)
- Conway-Maxwell Poisson regression (COM) (Conway and Maxwell 1962; Sellers and Shmueli 2010)
- NB-L (Lord, Guikema, and Geedipally 2007)
- Double Poisson (Efron 1986; Miranda 2013; Geedipally, Lord, and Dhavala 2013)

For those who are interested in following up on this class of models, the Bibliography has articles by the researchers listed. R software is available for the listed models as well, and Stata commands exist for the top two

(Harris, Hilbe, and Hardin 2013). However, to give a sense of how these types of models work, I'll outline the generalized Waring model. The model is essentially akin to generalized Poisson regression, but with an additional parameter.

The model is named after Edward Waring (1736–1798), who first proposed a distribution from which the currently parameterized distribution is derived. The distribution was advanced by Newbold (1927), and Irwin (1968) and Xekalaki (1983), who developed its present form. The PDF is defined by Rodríguez-Avi et al. (2009) as

$$\frac{\Gamma(\alpha+\rho)}{\Gamma(\alpha)\,\Gamma(\rho)}\nu^{(\alpha-1)}\,(1+\nu)^{-(\alpha+\rho)} \text{ with } \alpha,\rho \geq 0 \qquad (9.6)$$

The three parameters are α, ρ, and ν, which reflect the variability in the data. The PDF posits that the variability in the data can be partitioned into three additive components:

- randomness,
- proneness – internal differences between individuals,
- liability – the presence of outside sources of variability not included directly in the model.

Xekalaki (1983) developed the three components of variability for this distribution and was instrumental, along with Rodríguez-Avi et al. (2009), in developing an R package for estimation of a model based on the generalized Waring distribution. The model statistics (mean, variance, etc.) and examples of the use of this model can be found in Hilbe (2011). The Stata command that follows is from Harris, Hilbe, and Hardin (2013). The PDF appears different from equation (9.6) but is in fact the same. The zero-inflated version of the model is used here since the data being modeled are substantially overdispersed because of excessive zero counts. The response term, *docvis*, has far more zeros than expected due to its mean value of 3.1:

```
. global xvar "outwork age female married"
. zinbregw docvis $xvar, vce(robust) nolog inflate($xvar)
```

```
Zero-inflated gen neg binomial-W regression   Number of obs   =       3874
Regression link:                              Nonzero obs     =       2263
Inflation link : logit                        Zero obs        =       1611
                                              Wald chi2(4)    =      87.78
Log pseudolikelihood = -8211.698              Prob > chi2     =     0.0000
```

```
------------------------------------------------------------------------
              |              Robust
      docvis  |     Coef.   Std. Err.      z    P>|z|    [95% Conf. Interval]
--------------+---------------------------------------------------------
docvis        |
     outwork  |  .2033453   .0696412    2.92   0.004    .0668511   .3398396
         age  |  .0144888   .0025891    5.60   0.000    .0094143   .0195633
      female  |  .1779968   .0632358    2.81   0.005    .0540569   .3019366
     married  |   -.08596   .0719044   -1.20   0.232   -.2268901   .0549701
       _cons  |  .6758925   .1368147    4.94   0.000    .4077405   .9440444
--------------+---------------------------------------------------------
inflate       |
     outwork  | -.1881244   .1430329   -1.32   0.188   -.4684638    .092215
         age  | -.0253405   .0052679   -4.81   0.000   -.0356654  -.0150156
      female  | -.5850628   .1244194   -4.70   0.000   -.8289203  -.3412052
     married  |  .0060488   .1471382    0.04   0.967   -.2823369   .2944345
       _cons  |  .4636771   .2425741    1.91   0.056   -.0117594   .9391136
--------------+---------------------------------------------------------
   /lnrhom2   |  .1981729   .0867388                     .028168   .3681778
       /lnk   |  1.068476   .0799546                     .911768   1.225184
--------------+---------------------------------------------------------
         rho  |  3.219173   .1057496                    3.028568   3.445099
           k  |   2.91094   .2327431                    2.488719   3.404794
------------------------------------------------------------------------
Vuong test of zinbregw vs. gen neg binomial(W):   z =  0.92   Pr>z = 0.1777
```

k is the *proneness* parameter. For this model, it is a measure of a patient's likelihood of visiting a doctor again after earlier visits. k is defined as $\alpha\mu/(\rho-1)$. A k of 2.9 indicates that the model explains more variation than an NB2 negative binomial model on the same data. *Rho* (ρ) is the *liability* parameter, indicating unaccounted-for variation. There is considerable outside variation in the data.

The Vuong test advises the user as to the relationship of the zero-inflated model to a standard Waring model. If the *p*-value is significant and z positive, a zero-inflated model is preferred. A significant *p*-value with a negative *z*-statistic prefers the nonzero-inflated model. A nonsignificant *p*-value indicates no preference for either model. This is the case here.

For an R implementation, the **GWRM** package for the Waring regression model can be downloaded from CRAN. The same data as used for the Stata example are used here as well. Note that the R function does not allow for zero-inflated modeling. The results will differ a bit between the two models,

TABLE 9.10. R: Generalized Waring Regression

```
library(COUNT); library(GWRM)
data(rwm5yr); rwm1984 <- subset(rwm5yr, year==1984)
war <- GWRM.fit(docvis ~ outwork + age + female + married, data=rwm1984)
GWRM.display(war)
```

and more so for the variance parameters. (R code for the generalized Waring regression is given in Table 9.10.)

```
> GWRM.display(war)
$Table
        covars    estimates           se           z              p
1 (Intercept) -0.24507662 0.121905501  -2.010382 4.439078e-02
2     outwork  0.26839550 0.064308985   4.173530 2.999156e-05
3         age  0.02500018 0.002508417   9.966516 2.135897e-23
4      female  0.41875059 0.061392023   6.820928 9.045406e-12
5     married -0.10661269 0.068266458  -1.561714 1.183554e-01
$betaII
  par      coef     value
1   k -0.5201369 0.5944391
2  ro  1.0447706 3.8427462

$Fit
  log-likelihood       AIC       BIC   df
1     -8273.765 16561.53 16605.36 3867
```

The value of k, the proneness parameter, is .59, also indicating that it adjusted for variation in the data better than the negative binomial model. See Hilbe (2011) and Harris, Hilbe, and Hardin (2014) for a more extensive discussion. The generalized Famoye negative binomial model, Famoye and Smith's EPPM, and other models are discussed in the e-book *Extensions to Modeling Count Data*. Stata software for Famoye, zero-inflated Famoye, and truncated Famoye are available on the book's web site, with directions on their use. The same Waring models are available as well.

9.8 Bayesian Count Models – Future Directions of Modeling?

Bayesian estimation has become popular only in the last two decades. Actually, though, it is likely more accurate to state that it has become popular since the mid-2000s. There were certainly Bayesians before this, but the vast majority of statisticians adhered to what is called the frequentist interpretation of statistics. The statistical models we have thus far discussed in this book are based on a frequentist interpretation. Frequentists assume that there are fixed but unknown true parameters of a probability distribution that best describe the data we are modeling. The model data are regarded as a random sample of data from a greater population of data, which is distributed according to the parameters of a specific probability distribution. Again, the model parameters are fixed; the analyst's goal is to model the data so that we derive an estimate as close as possible to the true parameter values. Maximum likelihood estimation is the method used most often for estimating the parameters.

Bayesians, on the other hand, argue that parameters are themselves random variables, and each has its own distribution in a model. Each predictor coefficient is a random variable and can be entirely different from other coefficients in the model. We may initially assume that a binary variable is best described by a Bernoulli or a more general binomial distribution, a count described by a Poisson or negative binomial distribution, a continuous variable described by a normal or Gaussian distribution, or a positive-only continuous variable described by a gamma distribution. For a binary variable, for instance, we assume that it is best described by a Bernoulli distribution, which we call a posterior distribution. It is mathematically represented as a likelihood – not a log-likelihood.

A key feature of Bayesian statistics is that we may have information about the data that might bear on the parameter values. This is called prior information, and it is described mathematically as a distribution – a prior distribution. The posterior distribution for each predictor is updated by the respective prior distribution at each iteration in the overall estimation process. The likelihood is multiplied by the prior, resulting in an updated posterior distribution:

$$p(\theta|y) \propto L(\theta) \, \pi(\theta) \tag{9.7}$$

$p(\theta|y)$ is the posterior distribution that we believe describes the predictors, $L(\theta)$ is the likelihood of the distribution that makes the data most likely, and $\pi(\theta)$ is the prior distribution, which describes the prior information we believe bears on the final posterior distribution of the predictor.

The posterior distribution may be simple, if no priors are involved, or quite complex, when there is no analytic solution or when an analytic solution is extremely difficult to calculate. Since most real data entail complex

TABLE 9.11. R: Bayesian Poisson MCMC

```
library(COUNT); library(MCMCpack); data(medpar)
summary(poi <- glm(los ~ hmo + white + type2 + type3,
                 family=poisson, data=medpar))
confint.default(poi)
summary(poibayes <- MCMCpoisson(los ~ hmo + white + type2 + type3,
              burnin = 5000, mcmc = 100000, data=medpar))
```

distributions, Bayesian statisticians turned to simulation as a method of estimation and now base nearly all estimation algorithms on some variety of MCMC simulation algorithm. MCMC is an acronym meaning "Markov Chain Monte Carlo," which is a method of repeatedly sampling from the posterior until the updates do not change – until they are stable. This may take 100,000 to 1 million or more iterations.

The estimated parameter coefficient is taken as the mean (or sometimes median) of the posterior distribution. The standard error of the mean is the coefficient standard error, and the 95% quantiles are the "credible intervals," in contrast with the frequentist "confidence intervals."

There is a substantial theory underlying Bayesian modeling. Interestingly, when uniform priors, also called noninformative or vague priors, are given in a Bayesian model, the parameter estimates are often nearly identical to frequency-based maximum likelihood estimates of the model. The differences in coefficient values begin when informative priors are added to the posteriors.

I should mention that at times there is no clear idea what distributional shape a posterior distribution can take, especially if the model is multidimensional. In such situations, one just lets the MCMC algorithm search to find a stable posterior, for which the mean, standard errors, and credible intervals can be taken. Usually, the first 10,000 to 50,000 iterations are discarded until the searching becomes more directed. This is called the burn-in phase. We do this because it is not helpful to have distributional values far to the extreme when using the posterior distribution mean for the estimated parameter value (coefficient).

The **MCMCpack** package, which can be installed from CRAN, is an R package (see Table 9.11) that has a number of functions performing Bayesian inference using posterior simulation. As an example, I will first use the **MCMCpoisson** function to obtain Bayesian Poisson estimates of parameters of the **medpar** data:

```
#  GLM POISSON
Coefficients:
             Estimate   Std. Error z value   Pr(>|z|)
(Intercept)   2.33293    0.02721   85.744    < 2e-16 ***
hmo          -0.07155    0.02394   -2.988    0.00281 **
white        -0.15387    0.02741   -5.613    1.99e-08 ***
type2         0.22165    0.02105   10.529    < 2e-16 ***
type3         0.70948    0.02614   27.146    < 2e-16 ***

> confint.default(poi)
                  2.5 %        97.5 %
(Intercept)   2.2796061    2.38626006
hmo          -0.1184786   -0.02462001
white        -0.2075991   -0.10014302
type2         0.1803908    0.26291271
type3         0.6582513    0.76070204
```

Compare these results with parameter estimates based on simulation. The results are nearly identical. Note also that we discarded the first 5000 (burnin = 5000), which is given in the display that follows, with valid iterations used in estimation being from 5001 to 105,000; 100,000 valid iterations created the posterior distribution from which the mean, standard errors, and credible intervals were taken. This has always struck me as a validation of using these statistical methods, where two very different estimation processes result in nearly identical statistical values.

The nice thing about using Bayesian modeling is that highly complex mixed models can be estimated the same way. Using frequentist methods, such as maximum likelihood and quadrature, results in values similar to those obtained using the simulation method of MCMC:

```
#  BAYESIAN POISSON
Iterations = 5001:105000
Thinning interval = 1
Number of chains = 1
Sample size per chain = 1e+05
1. Empirical mean and standard deviation for each variable,
   plus standard error of the mean:
              Mean      SD     Naive SE   Time-series SE
(Intercept)  2.33266  0.02729 8.629e-05     0.0003650
hmo         -0.07172  0.02364 7.475e-05     0.0003381
white       -0.15383  0.02750 8.697e-05     0.0003764
type2        0.22174  0.02077 6.568e-05     0.0002661
type3        0.70914  0.02597 8.213e-05     0.0004182
```

```
2. Quantiles for each variable:

                 2.5%      25%      50%      75%     97.5%
(Intercept)    2.2789   2.31432  2.33289   2.3511   2.38545
hmo           -0.1183  -0.08756 -0.07163  -0.0558  -0.02563
white         -0.2073  -0.17254 -0.15395  -0.1352  -0.09988
type2          0.1801   0.20782  0.22194   0.2361   0.26160
type3          0.6578   0.69171  0.70929   0.7265   0.76031
```

There are two basic MCMC methods in common use at this time. The first, and first to be developed for this class of models, is the Metropolis–Hastings algorithm. The second is Gibbs sampling. Both have had numerous variations developed from the originals.

The foremost software packages used for Bayesian modeling are WinBUGS/OpenBUGS, JAGS, SAS **Genmod**, SAS **MCMC**, and MLwiN. Cytel's LogXact can also perform MCMC-based modeling. Analysts often use R directly to create Bayesian models, but it is much easier to employ software that has already-developed built-in simulation routines with error checks, graphics, and so forth. WinBUGS is popular and is the subject of numerous books, but the developers have stopped development and are putting their help and development tools into OpenBUGS. JAGS is an acronym for "Just Another Gibbs Sampler." It can be run through the R environment by using the **R2jags** package on CRAN.

SAS is also an excellent tool for Bayesian modeling. I used SAS's **Genmod** for displaying examples of Bayesian negative binomial models in Hilbe (2011). For those interested in Bayesian modeling and its relationship to frequency-based modeling, I recommend Zuur, Hilbe, and Ieno (2013). The book compares fully worked-out examples using both traditional frequency-based modeling and Bayesian modeling of both generalized linear models (GLMs) and generalized linear mixed models (GLMMs), including normal, logistic, beta-binomial, Poisson, negative binomial, and gamma regression. R and JAGS code are used throughout. I also recommend Hilbe and Robinson (2013), in which R code for writing a Metropolis–Hastings algorithm is provided in annotated form. JAGS code for the **medpar** model in this section is provided on the book's web site.

9.9 SUMMARY

This chapter has covered various more "advanced" count models used by analysts. However, I only provided a brief overview of the subject area. I have placed additional material for many of these models on the book's web site in an electronic book titled *Extensions to Modeling Count Data*. I intend to provide timely updates to the e-book at http://works.bepress.com/joseph_hilbe/.

I began the chapter discussing exact Poisson modeling, which is recommended when a dataset is small and unbalanced; e.g., for a highly unequal ratio of 1s and 0s in binary variables, highly unequal number of observations per level in categorical models, or highly skewed, bifurcated, or multimodal data for continuous predictors. The *p*-values are exact, not asymptotic.

Truncated and censored count models are important models to have in one's tool chest. Neither have been available in major commercial statistical packages. Limdep was the only software offering Poisson and negative binomial truncated and censored models until they were added to the R **gamlss** package (Stasinopoulos, Rigby, and Akantziliotou 2008) by Bob Rigby for Hilbe (2011). Stata's new **treg** command (Hardin and Hilbe 2014b) provides truncated models for Poisson, NB2, PIG, generalized Poisson, NB-P, and NB-F. Censored models for these distributions will be available in later 2014.

We also discussed finite mixture models, where we find that a count variable can be composed of counts from different data-generating mechanisms. I believe that this model will have increased use in the future as well.

A brief look at the nonparametric modeling of count data was given in Sections 9.4 and 9.5. Space prohibited a longer exposition. See Hilbe (2011) for more details, as well as extensions to MCD. I highly recommend Zuur (2012) for dealing with GAM analysis of count data. This chapter also examined longitudinal count models, three-parameter models, and the newer Bayesian modeling of count data that is only now becoming popular. I hope that the overview of count models provided in this book encourages you to read more about this class of models and to incorporate them into your research.

SAS Code

Additional SAS code for models, graphs, and tables is located at http://works.bepress.com/joseph_hilbe/.

My appreciation goes out to Sachin Sobale, team leader for statistical and SAS programming, and Ashik Chowdhury, senior biostatistician, both at Cytel Corporation, India, for their good help in preparing the majority of the SAS code in this appendix. My appreciation is also given to Yang Liu, Sam Houston State University, for providing the SAS code to generate observed versus predicted Poisson and NB2 tables and figures (final Appendix section code). I thank Valerie Troiano and Kuber Dekar of the Institute for Statistics Education (Statistics.com) for arranging Cytel's assistance in this project.

POISSON

```
/*To call the data from .csv (excel) file*/
proc import OUT= WORK.rwm5yr
        DATAFILE= "D:\sas code from R and stata\rwm5yr.csv"
        DBMS=CSV REPLACE;
    GETNAMES=YES;
    DATAROW=2;
run;

/*Select data for year = 1984*/
data rwm1984;
  set rwm5yr;
  where year = 1984;
run;
```

```
/*To get the output similar to STATA on page 42 */
proc freq data = rwm1984;
   tables docvis/outcum;
run;

/*To get the output similar to STATA on page 42 */
proc means data = rwm1984 n  mean std min max ;
   var docvis ;
run;


/*To get the output similar to STATA on page 42 */
data pois;
   x =  poisson(3.162881,0);
run;


proc print data = pois;
run;


/*To get the output similar to STATA on page 42 */
proc freq data = rwm1984;
   tables outwork/outcum;
run;

/*To get the output similar to STATA on page 42 */
proc means data = rwm1984 n mean std min max ;
   var age ;
   output out = summ (drop = _freq_ _type_) mean = meanage;
run;

/*To get the output similar to STATA on page 43 */
proc sql;
   select count (age) as total,count (distinct age) as distinct
   from rwm1984;
quit;

/*derive deviation from the mean*/
data cage;
   if _N_ = 1 then set summ;
   set rwm1984;
   cage = age - meanage;
run;
```

```
/*To get the output similar to STATA on page 43 */
proc genmod data = cage;
  model docvis = outwork cage / dist = poisson
                                link = log;
run;


/*Alternate approach to get the output similar to what we are getting
using proc genmod(previous model)*/
/*To get the output similar to STATA on page 43 */
proc nlmixed data = cage ;
  parms _cons=0 _outwork=0 _cage=0;
  xb=_cons+_outwork*outwork+_cage*cage;
  mu = exp(xb);
  model docvis ~ poisson(mu);
run;
```

POISSON WITH PEARSON DISPERSION SCALED SEs

```
/*Call data medpar from .csv file*/
proc import OUT= WORK.Medpar
          DATAFILE= "D:\sas code from R and stata\medpar.csv"
          DBMS=CSV REPLACE;
     GETNAMES=YES;
     DATAROW=2;
run;


/*Poisson with Pearson dispersion scaled SEs */
/*To get the output similar to STATA on page 73 */
proc genmod data = Medpar ;
 model Length_of_Stay =  HMO_readmit_ __White Urgent_Admit
                         Emergency_Admit  / dist = poi
                                            link = log
                                            scale = PEARSON;
run;
```

POISSON WITH ROBUST VARIANCE ESTIMATOR

```
/*Sort the data set*/
proc sort data = Medpar;
  by descending type;
run;
```

```
/*Robust SEs from page 84 */
proc genmod data = Medpar order = data;
   class  type Provider_number   ;
   model Length_of_Stay =  type HMO_readmit_ __White  / dist =
poi link = log;
   repeated subject =  Provider_number;
run;
```

POISSON WITH EXPOSURE (OFFSETS)

```
/*To create fasttrakg data as given on page 53   */
data fasttrakg;
  input die cases anterior hcabg killip kk1 kk2 kk3 kk4;
  datalines ;
  5  19   0 0 4 0 0 0 1
  10 83   0 0 3 0 0 1 0
  15 412  0 0 2 0 1 0 0
  28 1864 0 0 1 1 0 0 0
  1  1    0 1 4 0 0 0 1
  0  3    0 1 3 0 0 1 0
  1  18   0 1 2 0 1 0 0
  2  70   0 1 1 1 0 0 0
  10 28   1 0 4 0 0 0 1
  9  139  1 0 3 0 0 1 0
  39 443  1 0 2 0 1 0 0
  50 1374 1 0 1 1 0 0 0
  1  6    1 1 3 0 0 1 0
  3  16   1 1 2 0 1 0 0
  2  27   1 1 1 1 0 0 0
run;

/*data shown on page 53 */
proc print data= fasttrakg;
run;

/*Create offset variable*/
data fasttrakg;
  set fasttrakg;
  lncase = log(cases);
run;
```

```
/*To get the output similar to STATA on page 53 */
ods output parameterestimates = estimate;
proc genmod data = fasttrakg ;
  model die = anterior hcabg  kk2 kk3 kk4 /   dist = poisson
                                             link = log
                                          offset = lncase;
run;


/*To get the actual result from the preceding model by
exponentiating the estimates*/
data fast;
  length parameter $15 IRR 8 StdErr 8 lowerci 8 upperci 8
ProbChiSq 8;
  set estimate;
  where parameter ne 'Scale';
  IRR = exp(estimate);
  lowerci = exp(LowerWaldCL);
  upperci = exp(UpperWaldCL);
  StdErr=sqrt(exp(estimate)**2*exp(StdErr**2)*(exp(StdErr**2) - 1));
  keep parameter IRR StdErr lowerci upperci ProbChiSq;
run;


/*To get the output similar to STATA on page 53 */
proc print data = fast;
run;
```

POISSON WITH MARGINAL EFFECTS (BOTH AVERAGE ME AND ME AT THE MEAN)

```
/* To get an estimate similar to STATA on page 56 using rwm1984 data*/
ods output parameterestimates = estimate;
proc genmod data = rwm1984 ;
   model docvis =  outwork age/ dist = poi
                                link = log;
run;


/*Take the average of age and outwork*/
proc means data = rwm1984 noprint;
  var outwork age;
  output out = summ(drop = _type_ _freq_) mean = outwork age;
run;
```

```
proc transpose data = summ out = trns (rename = (_name_ =
parameter col1 = mean));
   var outwork age;
run;

proc sort data = estimate;
   by parameter;
run;

proc sort data = trns;
   by parameter;
run;

data full;
   merge estimate trns;
   by parameter;
   where parameter ne 'Scale';
   if mean ne . then
   coeff = mean*estimate;
   else if mean = . then coeff = estimate;
run;

proc univariate data = full noprint;
   var coeff;
   output out = total sum = xb;
run;

data full1;
   set total; set full (where = (parameter = 'age'));
   dfdxb = exp(xb)*estimate;
run;

/* To get the output similar to STATA on page 56 */
/*Marginal Effects at mean   */
proc print data = full1;
   var dfdxb;
run;

/*Take the average of age and outwork*/
proc means data = rwm1984 noprint;
   var  docvis;
   output out = summ_(drop = _type_ _freq_) mean =  docvis;
run;
```

```
data full_;
  set summ_; set full (where = (parameter = 'age'));
  dfdxb_avg = docvis*estimate;
run;

/* To get the output similar to STATA on page 57 */
/*Average marginal effects*/
proc print data = full_;
  var dfdxb_avg;
run;
```

NB2 – NEGATIVE BINOMIAL (TRADITIONAL)

```
/*NB2 output on page 104 using rwm1984 data*/
proc genmod data = rwm1984 ;
  class id;
  model docvis =  outwork age female married edlevel2 edlevel3 edlevel4 /
dist = nb;
  repeated subject = id;
run;
```

ZERO-INFLATED POISSON

```
/*To get the output of STATA on page 145 */
/*To get the IRR and corresponding CIs exponentiate the estimate and
CIs derived from the following model*/
proc genmod data = rwm1984;
  model docvis = outwork age /dist=zip;
  zeromodel outwork age /link = logit ; /*zeromodel option is
available in SAS 9.2 or later version*/
run;
```

ZERO-INFLATED NEGATIVE BINOMIAL

```
/*To get the output of STATA on page 147 */
/*To get the IRR and corresponding CIs exponentiate the estimate
and CIs derived from the following model*/
proc genmod data = rwm1984;
  model docvis = outwork age /dist=zinb;
  zeromodel outwork age /link = logit ; /*zeromodel option is available
in SAS 9.2 or later version*/
run;
```

ZERO-TRUNCATED POISSON

```
/*Zero truncated poisson output of STATA on page 129 */
proc nlmixed data=Medpar;
    xb = intercept + hmo*HMO_readmit_+ white*__White
+Type2*Urgent_Admit
                        +Type3*Emergency_Admit;
    ll = Length_of_Stay*xb - exp(xb) - lgamma(Length_of_Stay + 1)
- log(1-exp(-exp(xb)));
    model Length_of_Stay ~ general(ll);
run;
```

ZERO-TRUNCATED NEGATIVE BINOMIAL

```
/*Zero truncated negative binomial output on page 130 */
proc nlmixed data=Medpar;
    xb = intercept + hmo*HMO_readmit_+ white*__Whit
+Type1*Elective_Admit;
    mu = exp(xb);
    m = 1/alpha;
    ll = lgamma(Length_of_Stay+m)-lgamma(Length_of_Stay+1)-lgamma(m) +
        Length_of_Stay*log(alpha*mu)-
(Length_of_Stay+m)*log(1+alpha*mu)
        - log(1 -( 1 + alpha*mu)**(-m));
    model Length_of_Stay ~ general(ll);
 run;
```

FINITE MIXTURE MODEL

```
/*To get the output similar to pages 164 and 165 using medpar data*/
/*To get the estimate for poisson distribution change dist=negbin
to dist = poisson in the following model */
proc fmm data=rwm1984; /*proc fmm is available in SAS 9.2 or later
version*/
    model docvis = outwork age /dist=negbin k = 2 cl;
run;
```

OBSERVED VERSUS PREDICTED COUNTS CHI2 GOODNESS-OF-FIT (0–20 VISITS)

```
/* Poisson Regression - Observed vs Expected Doctor Visits */
/*Directly open data set,then restore it under work library. These two
lines are not needed if import by clicking */
```

```
data rwm1984; set tmp3.rwm1984;
run;

/* Get observed counts */
proc freq data=rwm1984;
  table docvis / out=obs (rename=(count=observe));
run;

proc means data=obs;
  var observe;
  output out=sum1 sum=;
run;

/* Create a macro variable for number of observations */
data sum; set sum1;
  call symput("number",observe);
run;

/* Build Poisson model */
proc genmod data = rwm1984;
    model docvis = outwork age / dist=poisson;
    ods output ParameterEstimates=pe;
run;

* Transpose parameter estimate */
proc transpose data=pe out=estparms;
     var estimate; id parameter;
run;

/* Prepare for proc score step */
data estparms; set estparms;
     drop scale;
     _TYPE_="PARMS";
run;

/* Add intercept to data so Var is not needed in proc score */
data datacopy; set rwm1984;
     Intercept=1;
run;

/* Get scores for poisson process */
proc score data=datacopy score=estparms out=pred type=PARMS;
run;

/* Compute predicted probabilities of specified counts 0-20 */
data pred; set pred;
```

```
        do i= 0 to 20;
        if estimate=. then py=.;
        py=pdf('poisson',i,exp(estimate)); output; end;
        keep i py;
run;


/* Obtain the average of the above probabilities */
proc means data=pred;
        class i;
        output out=means mean(py)=;
run;


/* Obtain the expected (predicted) counts */
data expect; set means;
        drop _type_ _freq_;
        if i=. then delete;
        expect=py*&number;
run;


/* Join data sets and create table */
proc sql;
    create table exptobs as
    select i as count,observe,expect,observe-expect as diff
        from expect as l left join
            obs as r
        on l.i=r.docvis;
run;


/* Get proportion */
data prop; set exptobs;
        obsprop=observe/&number*100;
        preprop=expect/&number*100;
        diffprop=diff/&number*100;
        drop observe expect diff;
run;


/*Calculate the difference between observed and expected proportion
        and then divide the result by expected proportion (def. Chi
        square = sum((square(o-e))/e)*/
data count_; set prop;
  diff_sq = diffprop*diffprop;
  if preprop ne 0 then
  div = diff_sq/preprop;
run;
```

```
/* Taking summation of (observed-expected)**2/expected */
proc means data = count_ noprint;
  var div;
  output out = summ (drop = _freq_ _type_)sum = obs_chisq n = n;
run;

/* Calculate tabulated chi square value and p value */
data chi; set summ;
  alpha = 0.05;
  df = n-1;
  chi_table = cinv((1-alpha),df);
  p_value =put(1-probchi(obs_chisq, df), pvalue6.4);
run;

/* Print the Chi2 GOF statistic */
proc print data = chi;
run;

/* Print the table */
title1 h=2 justify=c "Table 3.6 Observed versus Predicted Counts";
proc print data=exptobs noobs;
run;
title1;

/* Plot the figure */
title2 h=2 justify=c "Figure 3.1 Observed vs Expected Doctor Visits";
axis1 label= (angle= 90 "Doctor Visits");
axis2 label= ("Number Visits to Physician");
legend1 label=none frame;
proc gplot data=exptobs;
     plot observe*count expect*count/overlay haxis=axis2 vaxis=axis1
legend=legend1;
     symbol1 interpol=join width=1.5 value=triangle c=steelblue;
     symbol2 interpol=join width=1.5 value=circle c=indigo;
run;

title2;
/* Negative Binomial Regression - Observed vs Expected Doctor Visits */
/* Build NB2 model */
proc genmod data = rwm1984;
     model docvis = outwork age / dist=negbin;
     ods output ParameterEstimates=pe_nb;
run;
```

```
/* Transpose parameter estimate */
proc transpose data=pe_nb out=estparms_nb;
     var estimate; id parameter;
run;

/* Create a macro variable for alpha */
data alpha; set estparms_nb;
     call symput("alpha",dispersion);
run;

/* Prepare for proc score step */
data estparms_nb2; set estparms_nb;
     _TYPE_="PARMS";
     drop dispersion;
run;

/* Add intercept to data so Var is not needed in proc score */
data datacopy_nb; set rwm1984;
     Intercept=1;
run;

/* Get scores for NB2 process */
proc score data=datacopy_nb score=estparms_nb2 out=pred_nb type=PARMS;
run;

/* Compute predicted probabilities of specified counts 0-20 */
data pred_nb; set pred_nb;
     do i= 0 to 20;
     if estimate=. then py=.;
     py=pdf('NEGBINOMIAL',i,1/(1+&alpha*exp(estimate)),1/&alpha);
      output; end;
     keep i py;
run;

/* Obtain the average of the above probabilities */
proc means data=pred_nb;
     class i;
     output out=means_nb mean(py)=;
run;

/* Obtain the expected (predicted) counts */
data expect_nb; set means_nb;
    drop _type_ _freq_;
```

```
         if i=. then delete;
         expect=py*&number;
run;

/* Join data sets and create table */
proc sql;
   create table exptobs_nb as
   select i as count,observe,expect,observe-expect as diff
       from expect_nb as l left join
           obs as r
       on l.i=r.docvis;
run;

/* Get proportion */
data prop_nb; set exptobs_nb;
       obsprop=observe/&number*100;
       preprop=expect/&number*100;
       diffprop=diff/&number*100;
       drop observe expect diff;
run;

/*Calculate the difference between observed and expected proportion
       and then divide the result by expected proportion (def. Chi
       square = sum((square(o-e))/e)*/
data count_nb; set prop_nb;
  diff_sq = diffprop*diffprop;
  if preprop ne 0 then
  div = diff_sq/preprop;
run;

/* Taking summation of (observed-expected)**2/expected */
proc means data = count_nb noprint;
  var div;
  output out = summ_nb (drop = _freq_ _type_)sum = obs_chisq n = n;
run;

/* Calculate tabulated chi square value and p value */
data chi_nb; set summ_nb;
  alpha = 0.05;
  df = n-1;
  chi_table = cinv((1-alpha),df);
  p_value =put(1-probchi(obs_chisq, df), pvalue6.4);
run;
```

```
/* Print the Chi2 GOF statistic */
proc print data = chi_nb;
run;

/* Print the table */
title3 h=2 justify=c "Table 3.7 Observed versus Predicted Counts-NB2";
proc print data=exptobs_nb noobs;
run;
title3;

/* Plot the figure */
title4 h=2 justify=c "Figure 3.2 Observed vs Expected Doctor
    Visits-NB2";
axis1 label= (angle= 90 "Doctor Visits");
axis2 label= ("Number Visits to Physician");
legend1 label=none frame;
proc gplot data=exptobs_nb;
     plot observe*count expect*count/overlay haxis=axis2 vaxis=axis1
legend=legend1;
     symbol1 interpol=join width=1.5 value=triangle c=steelblue;
     symbol2 interpol=join width=1.5 value=circle c=indigo;
run;
title4;
```

CENSORED POISSON MODEL

```
/* Hilbe, Censored Poisson, cenpois.sas. See Book web site
```

Bibliography

Akaike, H. 1973. "Information Theory and Extension of the Maximum Likelihood Principle," in *Second International Symposium on Information Theory*, ed. B. N. Petrov and F. Csaki, pp. 267–281. Budapest: Akademiai Kiado.

Amemiya, T. 1984. "Tobit Models: A Survey." *Journal of Econometrics* 24: 3–61.

Anscombe, F. J. 1953. "Contribution to the Discussion of H. Hotelling's Paper." *Journal of the Royal Statistical SocietySeries B* 15 (no. 1): 229–230.

Bailey, M., M. A. Collins, J. D. M. Gordon, A. F. Zuur, and I. G. Priede. 2008. "Long-term Changes in Deep-water Fish Populations in the North East Atlantic: A Deeper-Reaching Effect of Fisheries?" *Proceedings of the Royal Society B* 275: 1965–1969.

Barnett, A., N. Koper, A. Dobson, F. Schmiegelow, and M. Manseau. 2010. "Using Information Criteria to Select the Correct VarianceENCovariance Structure for Longitudinal Data in Ecology." *Methods in Ecology and Evolution* 1 (no. 1): 15–24.

Cameron, A. C., and P. K. Trivedi (1998). *Regression Analysis of Count Data*. Cambridge: Cambridge University Press.

Carroll, R. J., and D. Ruppert. 1981. "On Prediction and the Power Transformation Family." *Biometrika* 68: 609–615.

Consul, P. C. 1989. *Generalized Poisson Distributions: Properties and Applications*. New York: Marcel Dekker.

Conway, R. W., and W. L. Maxwell. 1962. "A Queuing Model with State Dependent Service Rates." *Journal of Industrial Engineering* 12: 132–136.

Cragg, J. C. 1971. "Some Statistical Models for Limited Dependent Variables with Application to the Demand for Durable Goods." *Econometrica* 39: 829–844.

Dean, C., and J. F. Lawless. 1989. "Tests for Detecting Overdispersion in Poisson Regression Models." *Journal of the American Statistical Association* 84: 467–472.

Desmarais, B. A., and J. J. Harden. 2013. "Testing for Zero Inflation in Count Models: Bias Correction for the Vuong Test." *Stata Journal* **13** (no. 4): 810–835.

Dohoo, I., W. Martin, and H. Stryhn. 2012. *Methods in Epidemiologic Research*. Charlottetown:VER Publishing.

Efron, B. 1986. "Double Exponential Families and Their Use in Generalized Linear Regression." *Journal of the American Statistical Association* **81**: 709–721.

Ellis, P. D. (2010). *The Essential Guide to Effect Sizes*. Cambridge: Cambridge University Press.

Faddy, M., and D. Smith. 2012. "Analysis of Count Data with Covariate Dependence in Both Mean and Variance." *Journal of Applied Statistics* **38**: 2683–2694.

Famoye, F. 1993. "Restricted Generalized Poisson Regression Model." *Communications in Statistics, Theory and Methods* **22**: 1335–1354.

    and K. Singh. 2006. "Zero-Truncated Generalized Poisson Regression Model with an Application to Domestic Violence." *Journal of Data Science* **4**: 117–130.

Fabermacher, H. 2011. "Estimation of Hurdle Models for Overdispersed Count Data." *Stata Journal* **11** (no. 1): 82–94.

    2013. "Extensions of Hurdle Models for Overdispersed Count Data." *Health Economics* **22** (no.11): 1398–1404.

Faraway, J. J. 2006. *Extending the Linear Model with R*. Boca Raton, FL: Chapman & Hall/CRC.

Flaherty, S., G. Patenaude, A. Close, and P. W. W. Lutz. 2012. "The Impact of Forest Stand Structure on Red Squirrel Habitat Use." *Forestry* **85**: 437–444.

Geedipally, S. R., D. Lord, and S. S. Dhavala. 2013. "A Caution about Using Deviance Information Criterion While Modeling Traffic Crashes." Unpublished manuscript.

Gelman, A., and J. Hill. 2007. *Data Analysis Using Regression and Multilevel/ Hierarchical Models*. Cambridge: Cambridge University Press.

Goldberger, A. S. 1983. "Abnormal Selection Bias," in *Studies in Econometrics, Time Series, and Multivariate Statistics*," ed. S. Karlin, T. Amemiya, and L. A. Goodman, pp. 67–85. New York: Academic Press.

Greene, W. H. 2003. *Econometric Analysis*, fifth edition. New York: Macmillan.

    2006. *LIMDEP Econometric Modeling Guide, Version 9*. Plainview, NY: Econometric Software Inc.

    2008. Functional Forms for the Negative Binomial Model for Count Data, *Economics Letters*, **99** (no. 3): 585–590.

Hamilton, L. C. 2013. *Statistics with Stata, Version 12*. Boston: Brooks–Cole.

Hannan, E. J., and B. G. Quinn. 1979. "The Determination of the Order of an Autoregression." *Journal of the Royal Statistical Society Series B* **41**: 190–195.

Hardin, J. W. 2003. "The Sandwich Estimate of Variance," in *Maximum Likelihood of Mis-specified Models: Twenty Years Later*, ed. T. Fomby and C. Hill, pp. 45–73. Elsevier: Amsterdam.

and J. M. Hilbe. 2013a. *Generalized Linear Models and Extensions*, third edition. College Station, TX: Stata Press/CRC.

and J. M. Hilbe. 2013b. *Generalized Estimating Equations*, second edition. Boca Raton, FL: Chapman & Hall/CRC.

and J. M. Hilbe. 2014a. "Regression Models for Count Data Based on the Negative Binomial(p) Distribution." *Stata Journal* 14.

and J. M. Hilbe. 2014b. "Truncated Regression Models for Count Data." *Stata Journal* 14.

Harris, T., J. M. Hilbe, and J. W. Hardin. 2013. "Modeling Count Data with Generalized Distributions." *Stata Journal*.

Harris, T., Z. Yang, and J. W. Hardin. 2012. "Modeling Underdispersed Count Data with Generalized Poisson Regression." *Stata Journal* 12 (no. 4): 736–747.

Hastie, T., and R. Tibshirani. 1986. "Generalized Additive Models." *Statistical Science* 1 (no. 3): 297–318.

1990. *Generalized Additive Models*. New York: Chapman & Hall.

Hausman, J. A. 1978. "Specification Tests in Econometrics." *Econometrica* 46: 1251–1271.

B. Hall, and Z. Griliches. 1984. "Econometric Models for Count Data with an Application to the Patents–R&D Relationship." *Econometrica* 52: 909–938.

Heckman, J. 1979. "Sample Selection Bias as a Specification Error." *Econometrica* 47: 153–161.

Heilbron, D. 1989. "Generalized Linear Models for Altered Zero Probabilities and Overdispersion in Count Data." Technical Report, Department of Epidemiology and Biostatistics, University of California, San Francisco.

Hilbe, J. M. 1993a. "Generalized Linear Models." *Stata Technical Bulletin* 11: sg16.

1993b. "Generalized Linear Models Using Power Links." *Stata Technical Bulletin* 12: sg16.1.

1993c. "Log Negative Binomial Regression as a Generalized Linear Model." Technical Report COS 93/94–5–26, Department of Sociology, Arizona State University.

1994a. "Negative Binomial Reegression." *Stata Technical Bulletin* 18: sg16.5.

1994b. "Generalized Linear Models." *The American Statistician* 48 (no. 3): 255–265.

1998. "Right, Left, and Uncensored Poisson Regression." *Stata Technical Bulletin* 46: 18–20.

2000. "Two-Parameter log-gamma and log-inverse Gaussian Models," in *Stata Technical Bulletin Reprints*, pp.118–121. College Station, TX: Stata Press.

2005a. "CPOISSON: Stata Module to Estimate Censored Poisson Regression." Boston College of Economics, Statistical Software Components, http://ideas.repec.org/c/boc/bocode/s456411.html.

2005b. "CENSORNB: Stata Module to Estimate Censored Negative Binomial Regression as Survival Model." Boston College of Economics, Statistical Software Components, http://ideas.repec.org/c/boc/bocode/s456508.html.

2007a. *Negative Binomial Regression*. Cambridge: Cambridge University Press.

2007b. "The Co-evolution of Statistics and Hz," in *Real Data Analysis*, ed. S. S. Sawilowsky, pp. 3–20. Charlotte, NC: Information Age Publishing.

2009a. *Logistic Regression Models*. Boca Raton, FL: Chapman & Hall/CRC.

2009b. "CPOISSONE: Stata Module to Estimate Censored Poisson Regression (Econometric Parameterization)." Boston College of Economics, Statistical Software Components, http://ideas.repec.org/c/boc/bocode/s457079.html.

2009c. *Solutions Manual for Logistic Regression Models*. Boca Raton, FL: Chapman & Hall/CRC.

2010a. "Modeling Count Data," in *International Encyclopedia of Statistical Science*, ed. M. Lovric. New York: Springer.

2010b. "Generalized Linear Models," in *International Encyclopedia of Statistical Science*, ed. M. Lovric. New York: Springer.

2011. *Negative Binomial Regression*, second edition. Cambridge: Cambridge University Press.

2012. *Astrostatistical Challenges for the New Astronomy*. New York: Springer.

and A. P. Robinson. 2013. *Methods of Statistical Model Estimation*. Boca Raton, FL: Chapman & Hall/CRC.

Hilbe, J. M., and W. H. Greene. 2008. "Count Response Regression Models," in *Handbook of Statistics*, vol. 27, ed. C. R. Rao, J. P. Miller, and D. C. Rao, pp. 210–252. Amsterdam: Elsevier.

Hin, L., and Y. Wang. 2008. "Working-Correlation-Structure Identification in Generalized Estimating Equations." *Statistics in Medicine* **28**: 642–658.

Hinde, J., and C. G. B. Demietrio. 1998. "Overdispersion: Models and Estimation." *Computational Statistics and Data Analysis* **27** (no. 2): 151–170.

Huber, P. J. 1964. "Robust Estimation of Location Parameter." *The Annals of Mathematical Statistics* **35** (no. 1).

1967. "The Behavior of Maximum Likelihood Estimates under Nonstandard Conditions," in *Proceedings of the Fifth Berkeley Symposium on Mathematical Statistics and Probability*, pp. 221–233. Berkeley: University of California Press.

Hurvich, C. M., and C. Tsai. 1989. "Regression and Time Series Model Selection in Small Samples." *Biometrika* **76** (no. 2): 297–307.

Irwin, J. O. 1968. "The Generalized Waring Distribution Applied to Accident Theory." *Journal of the Royal Statistical Society Series A* **131** (no. 2): 205–225.

Lawless, J. F. 1987. "Negative Binomial and Mixed Poisson Regression." *Canadian Journal of Statistics* **15** (no. 3): 209–225.

Leisch, F., and B. Gruen. 2010. *Flexmix: Flexible Mixture Modeling*. CRAN.

Long, J. S. 1997. *Regression Models for Categorical and Limited Dependent Variables.* Thousand Oaks, CA: Sage Publications.

and J. Freese. 2006. *Regression Models for Categorical Dependent Variables Using Stata*, second edition. College Station, TX: Stata Press.

Lord, D., S. E. Guikema, and S. R. Geedipally. 2007. "Application of the Conway-Maxwell-Poisson Generalized Linear Model for Analyzing Motor Vehicle Crashes." Unpublished manuscript.

Machado, J., and J. M. C. Santos Silva. 2005. "Quantiles for Counts." *Journal of the American Statistical Association* **100**: 1226–1237.

Maindonald, J., and J. Braun. 2007. *Data Analysis and Graphics Using R*. Cambridge: Cambridge University Press.

McCullagh P. 1983. "Quasi-likelihood Functions." *Annals of Statistics* **11**: 59–67.

and J. A. Nelder. 1989. *Generalized Linear Models*, second edition. New York: Chapman & Hall.

Miranda, A. 2013. "Un modelo de valla doble para datos de conteo y su aplicación en el estudio de la fecundidad en México," in *Aplicaciones en Economía y Ciencias Sociales con Stata*, ed. A. Mendoza. College Station, TX: Stata Press.

Morel, J. G., and N. K. Neerchal. 2012. *Overdispersion Models in SAS*. Cary, NC: SAS Press.

Muenchen, R., and J. M. Hilbe. 2010. *R for Stata Users*. New York: Springer.

Mullahy, J. 1986. "Specification and Testing of Some Modified Count Data Models." *Journal of Econometrics* **33**: 341–365.

Nelder, J., and D. Pregibon. 1987. "An Extended Quasi-likelihood Function." *Biometrika* **74**: 221–232.

Newbold, E. M. 1927. "Practical Applications of the Statistics of Repeated Events, Particularly to Industrial Accidents." *Journal of the Royal Statistical Society* **90**: 487–547.

Pan, W. 2001. "Akaike's Information Criterion in Generalized Estimating Equations." *Biometrics* **57**: 120–125.

Rabe-Hesketh, S., and A. Skrondal. 2005. *Multilevel and Longitudinal Modeling Using Stata*. College Station, TX: Stata Press.

and M. Stasinopoulos. 2008. "A Flexible Regression Approach Using GAMLSS in R." Handout for a short course in GAMLSS given at International Workshop of Statistical Modelling, University of Utrecht.

Rodríguez-Avi, J., A. Conde-Sánchez, A. J. Sáez-Castillo, M. J. Olmo-Jiménez, and A. M. Martínez-Rodríguez. 2009. "A Generalized Waring Regression Model for Count Data." *Computational Statistics and Data Analysis* 53 (no. 10): 3717–3725.

Rouse, D. M. 2005. "Estimation of Finite Mixture Models." Masters Thesis, Department of Electrical Engineering, North Carolina State University.

Schwarz, G. E. 1978. "Estimating the Dimension of a Model." *Annals of Statistics* 6 (no. 2): 461–464.

Sellers, K. F., S. Borle, and G. Shmueli. 2012. "The COM-Poisson Model for Count Data: A Survey of Methods and Applications." *Applied Stochastic Models in Business and Industry* 28 (no. 2).

Sellers, K. F., and G. Shmueli. 2010. "A Flexible Regression Model for Count Data." *Annals of Applied Statistics* 4 (no. 2): 943–961.

Shults, J., and J. M. Hilbe. 2014. *Quasi-Least Squares Regression*. Boca Raton, FL: Chapman & Hall/CRC.

Smith, D. M., and M. J. Faddy. "Mean and Variance Modelling of Under- and Overdispersed Count Data." *Journal of Statistical Software.*

EPPM.functions Counts. CRAN.

Smithson, M., and E. C. Merkle. 2014. *Generalized Linear Models for Categorical and Continuous Limited Dependent Variables*. Boca Raton, FL: Chapman & Hall/CRC.

Stasinopoulos, M., B. Rigby, and C. Akantziliotou. 2008. *Instructions on How to Use the gamlss Package in R*, second edition. CRAN.

Stone, C. S. 1985. "Additive Regression and Other Nonparametric Models." *Annals of Statistics* 13 (no. 2): 689–705.

Tang, W., H. He, and X. M. Tu. 2013. *Applied Categorical and Count Data Analysis*. Boca Raton, FL: Chapman & Hall/CRC.

Tutz, G. 2012. *Regression for Categorical Data*. Cambridge: Cambridge University Press.

Vickers, A. 2010. *What Is a P-Value Anyway?* Boston: Addison-Wesley.

Vittinghoff, E., and C. E. McCulloch. 2006. "Relaxing the Rule of Ten Events per Variable in Logistic and Cox Regression." *American Journal of Epidemiology* 165 (no. 6): 710–718.

Vuong, Q. H. 1989. "Likelihood Ratio Tests for Model Selection and Non-nested Hypotheses." *Econometrica* 57: 307–333.

Wang, Z. 2000. "Sequential and Drop One Term Likelihood-Ratio Tests." *Stata Technical Bulletin* 54: sg133.

Wedderburn, R. W. M. 1974. "Quasi-likelihood Functions, Generalized Linear Models and the Gauss–Newton Method." *Biometrika* **61**: 439–447.

Westfall, P. H., and K. S. S. Henning. 2013. *Understanding Advanced Statistical Methods*. Boca Raton, FL: Chapman & Hall/CRC.

White, H. 1980. "A Heteroskedasticity-Consistent Covariance Matrix Estimator and a Direct Test for Heteroskedasticity." *Econometrica* **48** (no. 4): 817–838.

Winkelmann, R. 2008. *Econometric Analysis of Count Data*, 5th ed. New York: Springer.

Xekalaki, E. 1983. "The Univariate Generalized Waring Distribution in Relation to Accident Theory: Proneness, Spells or Contagion?" *Biometrics* **39** (no. 3): 39–47.

Zhu, F. 2012. "Modeling Time Series of Counts with COM-Poisson INGARCH Models." *Mathematical and Computer Modelling* **56** (no. 9): 191–203.

Zou, Y., S. R. Geedipally, and D. Lord. 2013. "Evaluating the Double Poisson Generalized Linear Model." Unpublished manuscript.

Zuur, A. 2012. *A Beginner's Guide to Generalized Additive Models with R*. Newburgh: Highlands Statistics.

Zuur, A. F., J. M. Hilbe, and E. N. Ieno. 2013. *A Beginners Guide to GLM and GLMM with R: A Frequentist and Bayesian Perspective for Ecologists*. Newburgh: Highlands Statistics.

Zuur, A. F., A. A. Sveliev, and E. N. Ieno. 2012. *Zero Inflated Models and Generalized Linear Mixed Models with R*. Newburgh: Highlands Statistics.

Index